About the authors

Joanne Ruthsatz is an Assistant Professor of Psychology at Ohio State University. Her research has been published in distinguished scientific journals and she has been featured in various publications including *Time, Scientific American, Psychology Today* and *The Huffington Post*.

Kimberly Stephens is a freelance writer, and Joanne Ruthsatz' daughter. She has degrees from Princeton and Harvard Law School, and is currently a PhD candidate at Brown University.

The Prodigy's Cousin

The Prodigy's Cousin

The Family Link Between Autism
and Extraordinary Talent

JOANNE RUTHSATZ
AND
KIMBERLY STEPHENS

RIDER • SYDNEY • AUCKLAND • JOHANNESBURG

1 3 5 7 9 10 8 6 4 2

Rider, an imprint of Ebury Publishing,
20 Vauxhall Bridge Road,
London SW1V 2SA

Rider is part of the Penguin Random House group
of companies whose addresses can be found at
global.penguinrandomhouse.com

First published in Great Britain by Rider in 2016
Published in the United States by Current, an imprint of
Penguin Random House LLC, New York

www.penguin.co.uk

A CIP catalogue record for this book is available from the British Library

ISBN 9781846045257

Printed and bound in Great Britain by Clays Ltd, St Ives PLC
Designed by Leonard Telesca

Penguin Random House is committed to a sustainable future
for our business, our readers and our planet. This book is
made from Forest Stewardship Council® certified paper.

To Jim
—J.R.

To Dan
—K.S.

To call specific behavioral superiority a "mystery" is merely to sugarcoat scientific neglect.

—Ogden R. Lindsley, 1965

Contents

Contents

Introduction

This story begins on the back roads of a swampy Louisiana town. That's where, in 1998, a young couple were raising their six-year-old son, a child with round cheeks, thin lips, and a peculiar knowledge of jazz musicians.

That spring, Joanne Ruthsatz, then a psychology graduate student, took a thirty-some-hour train ride from Sandusky, Ohio, to New Orleans, rented a car, and drove through the bayou to the couple's small clapboard home.

She'd come all that way to see the boy—Garrett James.* He looked like a typical little kid—medium build, towheaded, and light-eyed. He loved trucks, spoke with a southern drawl, and listened politely when his parents introduced him to "Miss Joanne."

But he was definitely not typical. As a toddler, Garrett had crafted musical instruments out of household goods—spoons, keys, the vent in the wall—anything he could get his tiny hands on. His aunt gave him a toy guitar for his second birthday, and he stunned his parents when he used it to re-create songs he heard on the radio.

His parents bought him a real child-sized guitar a few months later. Garrett practiced during the day; he practiced at night. He scampered

---

* Throughout this book, we use the names and describe the lives of real children. Garrett James is an exception. Garrett's story is based on that of an actual child, but to protect his identity we have used a pseudonym and obscured the details of his development. The child's actual accomplishments are even greater than those described here—hard to imagine, we know!

to his instrument at every opportunity, held it while chatting with his parents, and strummed during breaks in the conversation. Music pulsed through the house; eventually, Garrett's parents asked their son to practice in the basement.

Garrett's love for music—his *need* for music—exploded from somewhere within him, charged through his fingertips, and burst into the world. At four years old, he performed in the family's yard, fronted a local band, and fielded invitations to play at music festivals and fairs. At one of these festivals, Garrett played in front of tens of thousands of fans. Physically, he was a speck on a sweeping stage; his famous adult co-star had to crouch down to approach eye level. When he played, though—a raucous, wholehearted number punctuated by foot tapping and hip swinging—he swallowed up the empty space. Little boy became music powerhouse.

Over the next two years, Garrett performed on a jazz album, in a movie, and on TV talk shows. All without ever having taken a formal music lesson, and all before his seventh birthday.

Garrett was a fascinating kid, but at that point Joanne's visit was a lark. She was within spitting distance of her Ph.D.; she just wanted to see if a theory she'd been working on could account for Garrett's abilities.

For most of Joanne's graduate school career, she'd been studying exceptional adult and teenage performers, trying to parse out what separates the successful from the less so. The nature versus nurture debate rubbed her the wrong way; surely both have a role to play in expertise. She had been working on a new theory, one based on the idea that at least *three* factors have a role in success: general intelligence, practice time, and skills specific to a particular field. Others had argued for the importance of each of these factors; it was the combination of the three that was novel.

She had already taken her theory for a test-drive and found some supporting evidence. College-level musicians scored higher on all three factors than high school musicians (who were presumably less expert):

they outwitted them on the IQ test, more successfully picked out changes in tone and rhythm (skills specific to their area of expertise), and had logged significantly more practice hours.

But could her theory account for the abilities of a child prodigy— one of those rare, preternaturally skilled, scientifically befuddling children who often outperform grown-up musicians, artists, and mathematicians? Joanne thought Garrett, an earnest, wide-eyed guitar phenom, would need an outlandish IQ and a masterful ear for music to make up for his relatively few years of practice.

Over the course of two days, Joanne gave Garrett an IQ test and a music aptitude test. Whenever there was a break, he bolted to his guitar to pick out a tune. Eventually, Garrett asked to go to McDonald's.

His timing wasn't great. Garrett still hadn't finished the memory section of the IQ test. But Joanne had three children of her own; she knew when to quit. She and Garrett, along with Garrett's mother, Sandra, drove to get lunch.

During the short trip, Joanne mulled over Garrett's test results. This kid was even more of a mystery than she had realized. Garrett had scored in the upper echelons of the music aptitude test, detecting changes in tone and rhythm with more accuracy than almost all of his age-mates, just as she had predicted he would. And he knocked the socks off the memory portion of the IQ test. He thundered through digit repetitions on his way to scoring in the ninety-ninth percentile for this section, despite having gotten tired of the testing and wrapping up early.

But the rest of Garrett's IQ-test results wasn't exactly what Joanne had expected. He did well, to be sure; there was no question that Garrett's general intelligence score was well above average. But it wasn't one in a million. He had a very high IQ, but it was nowhere near as exceptional as his abilities on the guitar.

Without a truly explosive IQ, how was he mastering music with such unbelievable speed?

At McDonald's, the trio ran into Sandra's sister, Susan, and Susan's son, Patrick. Garrett's mother introduced Joanne to the extended family. Joanne said hello to Susan; the two sisters talked. The teenage Patrick grunted, but he never said a word. He occasionally flapped his hands. As Sandra later told Joanne, Patrick was autistic.

Joanne's mind began to spin. *Did Garrett's talent have something to do with his cousin's autism?*

It was a peculiar idea. Garrett didn't *have* autism. He didn't appear to have any of the typical symptoms. You could have scoured the academic literature on child prodigies—what little there was of it in 1998—without finding any suggestion that autism might lie at the root of the kids' abilities.

But Joanne had seen it for herself. In the fluorescent light of a hamburger joint, two cousins, two biologically related children, stood side by side, one courted by the press for his musical prowess, the other struggling to master daily living skills.

By better understanding one child, could you help the other?

Fast-forward eighteen years, and Joanne has tracked down dozens of prodigies. To date, she's worked closely with more than thirty—the largest research sample of these rare children ever created. Their stories border on the fantastic: a two-year-old who loved to spell words like "algorithm" and "confederation"; a six-year-old painter captivated by Georgia O'Keeffe; a seven-year-old violinist with a powerful benevolent streak.

Joanne and her daughter, Kimberly Stephens, wrote the scholarly papers that stemmed from these encounters. But as the research advanced, they (hereinafter "we") realized that the relationship between prodigy and autism had implications that went far beyond an ivory-tower debate over the source of extraordinary talent.

Thus, this book. To explore the prodigies' lives and examine the underpinnings of their abilities, we draw on Joanne's years of research; dozens of interviews with the kids and their parents; media accounts of the children's lives; the medical records, education reports, and videos and photographs provided by the families; and previous child prodigy research. To flesh out how the prodigies' abilities connect with autism, we draw on hundreds of academic studies; the interviews we conducted with experts in the field; and extensive conversations with the families and friends of the autists we profile. It was an incredible amount of raw material, bursting with gems of information about extraordinary children, intrepid scientists, startling research findings, and fiercely devoted parents. We use it here to tell a scientific detective story of sorts in which we investigate what makes prodigies tick.

⟵

Child prodigy studies isn't a particularly crowded field. Despite perennial fascination with these children, research into the roots of their abilities is quite sparse. If there were a conference just for those who have firsthand experience working with prodigies, the organizers wouldn't rack up much of a catering bill; at best there would be a handful of attendees.

At least partly for this reason, child prodigies' abilities are a long-running mystery. How *can* a kid who still travels by car seat compose classical music or pontificate about astrophysics? But once you consider the possibility that the prodigies' startling skills might have something to do with autism, pieces start to fall into place. We'll begin by using the story of two brothers (one of whom counts corn kernels to check for Fibonacci numbers and plays Christmas carols *backward* on his ukulele for fun) to illustrate just how thin the line between prodigy and autism can be.

Then we'll follow the research trail as evidence of a potential link piles up. It's a journey that starts with the early struggle to sort out who qualifies as a child prodigy, which turns out to be a surprisingly complicated question. We'll dive into the prodigies' common attributes, such as extraordinary memories, eagle eyes for detail, and voracious appetites for their chosen subjects—many of which are also common attributes of autism. These traits shape the prodigies' lives, as we'll see in stories like that of the violinist who notices slight variations in the sounds of the subway chimes and the child physicist whose mind is a steel trap for numbers. We'll also take a close look at another, lesser-known characteristic of child prodigies—extreme empathy—and explore whether it is yet another trait that links prodigy to autism. Finally, we'll examine the emerging evidence that prodigy and autism may share genetic roots.

Along the way, we'll investigate the kids' family lives and the role the parents play in raising a child prodigy. We'll delve into the case of a typically developing teenager who became completely engrossed with music after he smacked his head against a church floor and explore whether his story means that such abilities exist somewhere inside us all. We'll dig into the evidence that suggests there are important underlying differences among the prodigies, distinct cognitive attributes that might prevent a math prodigy, for example, from becoming an art prodigy.

Toward the end, we'll explore the possibility—tenuous yet tantalizing—that studying child prodigies might advance our understanding of autism. It's a possibility with a lot of ifs (the genetic connection between the two would have to be confirmed, for starters). But if studying child wonders could potentially improve our understanding of autism, we suspect that a few more people might start showing up for those hypothetical child prodigy conferences.

⟶

Finally, a note on autism and the language we'll use in this story.

Autism varies a great deal in severity, and those diagnosed with autism (and their families) can face steep challenges. This is a book about the riddle presented by children with extreme abilities. We discuss autism research, but we do not focus on the serious difficulties that autists and their families can encounter, though this topic is clearly deserving of sustained and serious attention.

Changing perceptions of autism have led to some disagreement over whether autism should be considered a brain disorder that should be cured or whether it's more properly viewed as a unique way of seeing the world that should be valued and, when necessary, accommodated.

This relates to a split over terminology: Should someone diagnosed with autism be described as an "individual with autism" or as "an autistic"? Those who view autism as a disorder often prefer "individual with autism," while those who view autism as a distinct style of thinking and as part of an individual's identity are often partial to "an autistic" or "an autist" or "an autistic person."

We primarily use the term "autist" in this book, mostly because it provides a concise parallel to the term "prodigy," making, we hope, for easier reading. Similarly, we refer to autism and other conditions listed in the fifth edition of the *Diagnostic and Statistical Manual of Mental Disorders* (the *DSM-5*),* a book that is often referred to as psychiatry's bible, as "brain disorders" rather than "mental disorders." Following the lead of two prominent scientists at the National Institute of Mental Health (NIMH), we prefer this terminology because it better reflects these conditions' neural roots.

Before we get too far ahead of ourselves, though, let's start along a journey that, for Joanne, is more than eighteen years in the making.

Let's begin with a Canadian family of four, a doctor and an engineer whose sons gave them a firsthand look at the sometimes razor-thin edge between prodigy and autism.

---

* The World Health Organization publishes another diagnostic tool, the *International Classification of Diseases* (*ICD*). Both the *DSM* and the *ICD* are used in clinical practice and research. For the sake of clarity and simplicity, we'll focus our discussion on the *DSM*.

Chapter 1

A Warehouse of a Mind

Lucie* is well acquainted with *the look:* the raised eyebrows and slight drop of the jaw. It's often accompanied by a quick jab of silence that wounds the conversation. She used to encounter it a lot when she talked about her sons, so, for the most part, she didn't. She's a sociable, articulate French-Canadian doctor, but talking about her two boys is a little complicated; neither one is exactly typical.

By the time her older boy, Alex, was seven, he was a bright, wiry second grader with blue eyes and dark hair. He loved jujitsu, tickled his brother, and spent hours perfecting intricate Lego creations.

No one believed her when she said that he was—had been—autistic. She could *feel* the skepticism. There's no cure for autism, acquaintances said; you don't just snap out of it. Alex must have been misdiagnosed. It was too hard for them to reconcile the boy they knew—a well-liked, friendly kid who skillfully bombed down ski slopes—with autism.

Lucie knew better than to say much about her younger son, William, who was then a fidgety six-year-old with a cherubic face and a mischievous smile. He had also been diagnosed with autism, but his symptoms seemed to be melting away, just as his brother's had.

Then there was the matter of William's *hobbies:* letters and

---

* Lucie, Mike, Alex, and William wanted to share their story but still maintain their privacy. At their request, we're not disclosing their last names and have limited their location to Ontario, Canada. The facts of their story have not been altered.

numbers; math, science, and geography. He had taught himself the alphabet by eighteen months and read obsessively at two and a half. He delighted in math, could square numbers in his head by the time he was four, and tucked away people's birth dates so he could use them to calculate how old that person would be in a given year—past or future. He took a scorched-earth approach to reference materials like the atlas, the dictionary, and the periodic table. He consumed everything in his path; he kept it all in his "warehouse" of a mind.

It was a bit isolating for Lucie to keep so much to herself, especially when there was so very much to tell.

Lucie's first pregnancy was uneventful. She carried her baby to term and delivered a healthy boy. His eyes were pale blue and shaped like sideways teardrops; he had creamy skin, chubby cheeks, and a whisper of light brown hair. She and her husband, Mike, named their son Alexander. They called him Alex.

Within months, Lucie sensed that something was off. Alex was extremely—almost alarmingly—serious. Other babies laughed and gurgled, but never Alex. Nor did he form much of an attachment to her. When Lucie went back to her job as an anesthesiologist, six-month-old Alex was indifferent. There was no fussing when she left or excitement when she returned. As the months ticked by, he remained oblivious to her comings and goings.

He had a few quirks. Once he began to crawl, Alex discovered a crack in the grout between the tiles in the living room. He stared at it and traced it with his finger for long periods of time. He would venture off to another area of the room but then scamper back to the crack. He noticed whenever a lightbulb burned out, whether in his own home or a shopping center, and, just as with the crack in the grout, he would stare at it for long periods of time.

He wasn't speaking. Alex's first birthday passed, and he had not yet spoken his first word; he hadn't even begun to babble.

But Alex was quick to walk and run, and Lucie managed to ignore the worries that tugged at the back of her mind. The standard parenting books' refrain helped: every kid develops at his or her own pace.

Shortly after Alex turned one, they traveled to Pennsylvania to see family. The youngest of Lucie's nephews was Alex's age; the boys were born just a few weeks apart. The difference between the two was jarring: Alex's cousin gestured, followed his parents' gazes, waved, and pointed at things he wanted. He babbled as he played and cried when his parents left the room. Alex did none of those things.

When Alex was sixteen months old, Lucie delivered her second son, William. It was another uncomplicated pregnancy, another uncomplicated birth. She had another beautiful baby boy, an alert little guy with big eyes and a happy disposition. But this time when the doctor who came around to conduct newborn hearing tests offered Lucie the usual pamphlet on resources for children with language delays, she took one.

By that point there was no denying that something was going on with Alex. He showed no interest in his baby brother. He looked at him for a few seconds and then walked away. He was never affectionate with his parents, either. If they tried to hug him, Alex angled his entire body away to avoid contact. When Lucie tried to coax Alex into looking at her, he stared into the distance.

He covered his ears and screamed when he heard music play, bags crinkle, or water run. Lucie and Mike walked on eggshells: they quit listening to music; they only unpacked groceries when Alex was out of earshot; they waited until he was napping to rinse dishes— anything to avoid setting him off.

Though he had always seemed physically adept, Alex began doing an odd, repetitive dance when excited. He walked on his toes and flapped his hands.

He developed a fascination with drawers and doors. If Lucie or Mike left a door open, even the door to the dishwasher, he stared at it, silently, until someone closed it. He was fascinated by the sliding mechanism on which the drawers operated, and he opened and closed them repeatedly.

He had many toys, but he didn't play with them the way other children did. Often, he lined them up. He plucked small wooden trees off his train layout, positioned them carefully along the edge of the counter, and ran his eyes over them. Once the trees were perfectly aligned, he stood in front of the counter and rocked back and forth, staring at the configuration. He did the same thing with figurines; he did the same thing with cereal boxes.

A few weeks after William was born, Lucie took Alex to a drop-in clinic at First Words, the facility she had read about in her hospital pamphlet, the one that screened young children for speech and language delays. After Lucie filled out the forms with information about Alex, a speech and language pathologist pulled her into a private room. She was certain Lucie had made a mistake; she must have reversed the numerical rating scale. Lucie assured her that she had understood the scale correctly. "Her jaw dropped, and that's when my heart dropped," Lucie recalled.

The pathologist issued Alex an urgent referral for an appointment with a developmental pediatrician. Two months later, Alex was diagnosed with autism. He was only eighteen months old—young for such a diagnosis—but it wasn't even a close call. Alex checked nearly every box.

"We were devastated," Lucie recalled. "I was trying to be functional because I had a sixteen-month-old and a newborn. I was sleep deprived and trying to cope, but I was mourning the loss of everything I expected of my child. He may never speak; he may never show affection, go to school, make friends, get married, take care of himself, or have a job."

Getting help for Alex would be no easy task. The waiting list for publicly funded intensive behavioral therapy was roughly a hundred people long. It was a nonstarter: don't hold your breath, Lucie was advised; don't count on it.

Lucie was scrappy. She had worked her way through college;

afterward, five or six banks refused to lend her money for medical school, but Lucie kept filling out loan applications until she found a taker. She was trained in psychology and medicine, and Mike was an engineer. They could handle this. Lucie and Mike signed Alex up for speech therapy and occupational therapy while they waited for a spot in the behavioral therapy program. In the meantime, Lucie, who is meticulously organized, read everything she could find on autism intervention. She investigated evidence-based therapy, joined autism support groups, participated in workshops, and attended a three-day international autism conference.

She was intent on language. At every opportunity, she looked Alex in the face and enunciated words as clearly as she could, repeating them several times.

She got nowhere.

She worked on gestures. She tried to teach Alex to point to things he wanted, often while holding a colicky baby in her arms. When that didn't work, she tried following his gaze to see if she could figure out whether he was looking at a particular drink or toy, silently indicating his desires.

She got nowhere.

When Alex had another assessment a few months after his diagnosis, he was still yet to babble, let alone produce his first word. During the appointment, Alex paced the room and stared at the lights on the ceiling. He danced on his toes. He lay down on the floor and watched the wheels on a toy truck roll back and forth. He opened and closed cabinet doors. When someone stopped him from opening the cabinet doors, he cried and banged his head against the wall. His adaptive functioning, a measure meant to reflect a child's communication, daily living skills, socialization, and motor skills, was at the second percentile, indicating that Alex, who was a few months shy of two, functioned at the level of a ten-month-old.

Lucie threw out her parenting books. Alex wasn't meeting any of the developmental milestones. He wasn't even on the scale anymore.

The parenting tips the books provided for children Alex's age didn't apply. Reading them was just too depressing.

She did, eventually, have one breakthrough. At one of Alex's speech therapy sessions, Lucie had learned about the Picture Exchange Communication System (PECS). The idea was to put together a collection of pictures that someone who was nonverbal could use to indicate his needs. Lucie created a binder with dozens of pictures of food, milk, and toys and began slowly, painstakingly, introducing Alex to it.

Alex learned to find the right card and lay it on the table when he wanted something. After he mastered that step, Lucie taught him to put the card in her hand instead of on the table. She continued upping the ante until Alex learned to look her in the eyes when he handed her the card and then, finally, to cross the room to give Lucie the card.

It was a success. If Alex was thirsty, at least he could toss Lucie a picture of milk. But it was also demoralizing. It had taken almost a year for Alex to master using the binder. Was this as good as communication would get?

Lucie hired a private behavioral therapist. The therapist came to the family's home and immediately criticized the PECS binder: There were too many pictures on a page, she scolded. It was too overwhelming for Alex. The therapist's interactions with Alex were no better. Alex didn't respond well to her. When Lucie intervened, suggesting a couple of activities Alex enjoyed and mentioning how she thought he might be starting to demonstrate affection, the therapist told her that autistic kids didn't feel love. Lucie kicked her out of the house.

Despair, long circling, took hold. Lucie felt a slow but building tide of disappointment, then bitterness, then resentment. She was pouring hours and hours of time into trying to help her son, but the rewards of parenthood were elusive. Alex never hugged or kissed her. If she tried to hug or kiss him, he pushed her hand away. He never looked her in the eye or smiled.

When she was working at the hospital, she could sometimes, almost, shut everything out. She enjoyed what she did and felt effective. But when she got in her car at the end of the day, dread mounted. She didn't want to go home. "I felt hopelessly inadequate and ill equipped to deal with what had happened, and I was resentful deep down inside of Alex for making me feel that way," Lucie recalled. "Being resentful about your child, it makes you feel very guilty. It makes you feel like a monster."

Autism bloodied her marriage. She and Mike had always had a great relationship. They'd hit it off immediately when they met years before at a friend's birthday party and Mike, dressed in a black turtleneck and black-rimmed glasses, asked Lucie's opinion about the Canadian medical system. They had mutual friends; both loved Halloween; even planning the wedding was a breeze. But autism nearly ripped them apart.

Their finances were blown. They were a two-income household, a doctor and an engineer. They had never expected to pinch pennies, but as their bills mounted, they struggled to make ends meet.

They started couples therapy, and it helped. The therapist reassured them, over and over again, that their relationship was solid. All of the problems they were having, all of the stress, was completely circumstantial. They were big circumstances, but circumstances nonetheless. They could work through it.

⌐

When Alex was two, the family caught a break. The Portia Learning Centre, an organization that provides services for autistic children and children with developmental delays, opened a nearby location. Lucie and Mike sprinted through the door.

The therapists at the center began with an intensive assessment of Alex. They emphasized the importance of getting to know him, of studying his particular case. In a moment that Lucie will never forget, the therapist remarked on Alex's potential: I really see something in him, she said. Though other experts had told Lucie that Alex would never speak meaningfully, his new therapist said the first thing they were going to do was get him to talk.

Lucie told the therapist they had been working on speech for months. She told her how when she handed Alex milk, she would look directly at him and enunciate, as clearly as she could, "milk." Maybe that's too intense for him, the therapist suggested. Let's try something he might find a bit less overwhelming.

They started working with Alex in his home. During the first session, Alex stood in the living room in front of his model train set, slowly, quietly, moving the trains back and forth across the tracks. The therapist positioned herself behind him and watched him play. Slowly, quietly, she began making the same consonant sound over and over again: *mm mm mm mm mm mm.*

Back and forth, Alex moved the trains. *Mm mm mm mm mm mm.* Back and forth. *Mm mm mm mm mm mm.*

The ninetieth time the woman repeated the sound, Alex chimed in. *Mm mm mm mm mm mm.*

The therapist switched to a new consonant sound: *bb bb bb bb bb bb.*

Back and forth, Alex moved the trains. *Bb bb bb bb bb bb.* Back and forth. *Bb bb bb bb bb bb.*

The sixtieth time, Alex chimed in.

The therapist switched to a third consonant sound. This time, Alex chimed in on the fortieth repetition. "It was like, oh, is this all we needed to do?" Lucie remembered. "How hard have I been working on this at home? Why didn't anyone tell me to do this?"

But it was two steps forward, one step back. Once Alex seemed comfortable with his therapist, they tried switching his sessions to the learning center. Alex made it no farther than the entranceway, where he planted his feet and screamed. He reached for the light switch and began flicking it up and down, turning the lights on and off. The therapist interacted with him for a few sessions in the entranceway. Then she began putting toys and activities that interested him in front of the treatment room door. Eventually, she coaxed him inside.

Two months later, Lucie was feeding Alex lunch in his high chair while William napped. When Alex finished what she had given him, he stared at the bowl of fruit sitting on the counter.

At two and a half years old, Alex said his first word: "pear."

Lucie's heart thumped. She stopped breathing. She sat completely, perfectly still.

"Pear," Alex repeated.

Lucie sobbed. "Pear! Pear!" she cried out. She rushed to the counter and began slicing her son's favorite fruit. She sliced so quickly she nearly chopped her finger off.

Two and a half years of silence. Thousands of dollars in therapy. Periods of despair. And now, finally, "pear."

A wall had crumbled. Word followed word, and over the following weeks Alex built them into sentences. A few months later, he began using his words to ask for things.

Breaking the communication barrier seemed to poke holes in Alex's social isolation as well. He responded more. Over the next few months, he began playing and laughing with his brother. He no longer gazed off into the distance but looked right at other people. He smiled; he laughed. He let his therapist tickle him.

In February, just three months after Alex had said his first word, he played chase with his father. It was an interactive game, the kind autistic children weren't supposed to be able to play. Lucie watched, amazed, as Mike ran around the house, doing a loop through the kitchen and the living room, while Alex ran after him. Lucie grabbed her camera. Mike switched directions unexpectedly, and Alex laughed. He switched directions again, and Alex laughed. Mike stopped and turned, letting Alex catch up. Alex looked straight at his dad, ran into his arms, and hugged him. The two high-fived. Tears filled Lucie's bright blue eyes, and she had trouble focusing the camera.

At Portia, Alex began initiating social interactions with the other children. The therapists urged Lucie and Mike to get Alex into a different environment. The children at the center were predominantly autistic; few of them would engage with Alex socially. The therapists recommended a nearby nursery school they thought would be open to enrolling Alex and allowing him to have a therapist accompany him during the school day. "I was surprised when they suggested it, and really scared," Lucie recalled. "But I trusted her, the woman who runs Portia; I trusted her infinitely with anything because she was the one who was responsible for his progress. Whatever she would have said, I would have done."

There was improvement; there were bumps. Alex could communicate with his teachers, but he had no idea how to make friends. Lucie and Mike plotted with the teachers, searching for ways to help Alex integrate with his class.

Things got better. After a year, Alex enrolled in a small but mainstream, private junior kindergarten (the Ontario equivalent of pre-kindergarten). His therapist tagged along. She accompanied him in the classroom, helping him to integrate in group settings, and offered one-on-one teaching. There were still challenges. There were days when Alex ran around the classroom screaming and days when he threw toys. But eventually those behaviors began to fade. They lessened in intensity. They disappeared. Lucie and Mike began cutting back on the number of hours that Alex's therapist accompanied him at school. Eventually, Alex's teachers said she was no longer needed at all.

When Alex was four and a half years old, his eligibility for public services was reassessed. It was a mere two and a half years after his diagnosis, but the experience was completely different. During his first visit, Alex had made infrequent eye contact and never spoke or gestured. This time, he made direct eye contact, introduced himself to the examiner, and initiated conversation. During his first visit, he had paced the room, opened and closed cabinet doors, stared at the lights, and lined up toys. This time, he complied with the examiner's requests and played with the toys in a more typical manner. During his first visit, the examiner had placed his adaptive skills at the second percentile; this time, he was at the forty-fifth percentile—comfortably within the average range.

There were still a few remnants of Alex's autism. He had trouble using the restroom in unfamiliar places. He had a comfort toy he spoke to in a repetitive manner several times a day, and he occasionally walked on his toes. But bottom line: Alex no longer qualified as autistic. His diagnosis was stripped away. He could transition out of behavioral therapy and was ready for full integration into the school system.

Although Lucie and Mike had watched Alex's progress, it was still something of a shock. *Could a kid really grow out of autism?*

Those at Portia were thrilled; Alex's development was breathtaking. But for Lucie and Mike, it was hard to fully rejoice. Just over a year earlier, William, Lucie and Mike's second son, had received an autism diagnosis of his own.

—

Lucie had been in a hurry to have a second child.

"Basically, I'm up in stirrups, I push Alex out, and I ask how many weeks do I have to wait until I have another kid?" Lucie remembered.

Because she was hell-bent on making up time after putting off starting a family during her medical training, she was pregnant with William before she knew about Alex's autism. If she had known ahead of time, William would probably never have been born: "I would have been too scared to get pregnant again."

After Alex was diagnosed, Lucie had watched William like a hawk. She knew that if a boy had an autistic older sibling, the likelihood that he, too, would have autism shot through the roof.

At first, Lucie thought they had gotten lucky. William wasn't as serious as Alex had been. After a colicky first few months, he was a quiet but happy baby. He was much more interactive than Alex had been. His face lit up when he saw his parents; he was enthralled with his older brother. William's play was typical in a way that Alex's had never been. If you gave him toy cars, he didn't line them up; he pushed them back and forth. The boys are like night and day, Lucie thought. William is going to be fine.

In fact, Lucie began to suspect that William was more than fine. He might actually be quite bright. He had an almost intuitive knowledge of shapes. At twelve months old, he zipped through peg puzzles without any trial and error. He plunked the shapes—even the relatively complex ones—right into place.

He adored letters. At four months old, his face lit up when he looked at books, and at eight or nine months, he began pointing to the words. Whenever he could get his hands on Alex's PECS binder, he lay on the floor and traced the letters of words like "cookie," "milk," and "Magna Doodle" with his pudgy fingers. Once he started crawling, he ignored the assorted toys around the living room and motored himself toward the books. He picked them up, turned the pages, and spent long periods of time staring at them.

He was absorbing far more than Lucie and Mike realized. When William was eighteen months old, Lucie took him and Alex shopping. It was an activity both boys hated. After an hour or two, Lucie took them to the play area in the mall and let them loose.

A set of letters—randomly distributed—decorated the play area carpet. William was immediately drawn to them. He ignored all the kids and all the toys and stared at them. Lucie watched as William began moving from one letter to the next. He found the *A* and stood on top of it. Then the *B* and the *C* and the *D*. He kept progressing all the way to *Z*.

Lucie was a bit surprised. She hadn't been working on the alphabet with William. She hadn't been working on much of anything with him. He was almost always happy, almost always content, and at the time Lucie was in the thick of battling autism with Alex, who was then in his first year at Portia.

William started moving through the letters again, this time in a different—but still deliberate—progression. The hair on Lucie's arms stood up. William, her eighteen-month-old child, was going through the alphabet *backward*. Lucie didn't think *she* could recite the alphabet backward.

But knowing that William had memorized the alphabet only exacerbated the sense of unease that had taken hold in the pit of Lucie's stomach. Over the previous few months, Lucie had noticed changes in her second son.

William's interest in letters and numbers had darkened into an obsession. He took his books to the corner of the room and spent hours staring at them. His social interests began to fade. William grew increasingly upset when things were not just how he wanted them, and changes in schedule became difficult. Lucie waited for months for William to utter his first words, but her baby hit the one-year mark, then the fourteen-month mark, then the sixteen-month mark, and those words still didn't come.

Lucie knew that autism had different faces. Sometimes autistic tendencies were evident almost immediately, as they had been with Alex. Sometimes a child might appear to be developing normally but then seem to slowly slide backward into autism's grasp. That was what she feared was happening with William.

Lucie took William for an assessment when he was just over two years old. The doctor thought she might be right. During his visit, William made some eye contact, certainly more than Alex had, but he was inconsistent about it. He occasionally responded to his name, but at least as often he seemed oblivious to it. He walked on his toes as Alex had done, and he had a ritualistic dance he performed when he was excited. Just like Alex, William never engaged in pretend play. His use of gestures was delayed, and even at two he rarely pointed. He produced some words, but he primarily used them to label things, not to communicate. The doctor diagnosed William with autism.

For Lucie, the disappointment flooded back in. The bitterness returned. In some ways, this diagnosis was even harder than the first; for months, she had been sure that William was going to be fine.

She broke down during a car ride with a friend. The appointments. The therapy. The workshops. Lucie didn't think she had the energy to start over.

But at least this time the course was charted. She had a formula at her fingertips. She had already enrolled William at Portia. She added him to the speech therapy waiting list; she added him to the occupational therapy waiting list. She closed her eyes, held her breath, and hoped for another miracle.

＊

At Portia, William's speech improved, but his progress was not as dramatic as Alex's had been. It was tough to offer incentives because his interests shifted unexpectedly.

After seven months of therapy, he still didn't respond to his name; eye contact was still fleeting. During an assessment, he yelled when the examiner tried to join him in completing a puzzle and nearly cried when his alphabet letters were not perfectly arranged. When frustrated, he bit his hand or tapped his head. He echoed the words of the examiner and his mother, a phenomenon known as echolalia, and chanted phrases—"Today is Friday" or "Go press the O"—out of context.

Despite William's struggles, Lucie began getting—more and more frequently—the distinct, sometimes chilling sense that his intelligence was "out there."

The December after William enrolled at Portia, he became engrossed with a set of magnetic letters that Lucie and Mike had bought for Alex. In the midst of the holidays, William took the letters off their magnetic board and placed them on a brown leather ottoman. Lucie went to get her camera to snap a couple of pictures. When she returned, she found William where she had left him, still playing with the letters.

She took a closer look. At two years and three months, William had spelled out two words across the ottoman, "Mommy" and "Daddy." He was midway through his third word, "William."

Lucie watched, stunned, as William continued to work, his head of tousled brown hair weaving back and forth between the ottoman and a stockpile of letters. He announced each new addition. He drew out its pronunciation for several seconds, as if he didn't want to let it go, as if merely *saying* the name of the letter was a delight. Eventually, his childish voice lilted upward, warbled, and broke, and William moved on to the next magnet. When he ran out of letters, he flipped a *W* over to use as an *M;* he rotated a 7 to create an *L.*

Over the next few weeks, William used the letters to spell increasingly difficult words. A few days after watching his aunt write out "algorithm," William re-created the word with the magnet letters; he spelled the same word out loud at dinner, squirming in his high chair, sporting a bib with a lion on it. Lucie discovered him spelling out the months of the year—correctly and in sequence—with the magnet letters. There was no calendar anywhere in the house. "There was a constant stream of stuff that he would always do that made the hair stand up on the back of my neck. It gave me goose bumps," Lucie said. "It was almost creepy."

By two and a half, William could battle his way through early readers, carefully keeping his place with his finger and sounding out each new word. After that, he read all the time. He read books, he read newspaper ads, he read labels; he read at home, and he read on the beach. Whenever he received a gift, he carefully read every word of the card before opening the present.

About a month before his third birthday, William began writing out words on Alex's chalkboard. At first, it was an exercise in frustration. When a letter wasn't perfect, William screamed and frantically erased it. But when he printed one to his specifications—magic. "I did it!" he would cry out, the curls above his forehead bouncing as he ran to Lucie. Within days, he was spelling out "Mommy" and "Amanda" (the name of one of his behavioral therapists). His writing wasn't restricted to the chalkboard; he used Lego bricks to craft words, piecing them together to spell out "Nana," and he wrote out the alphabet on graph paper.

William discovered the computer before he turned three and became mesmerized by fonts. He spent hours typing words in different fonts and developed opinions about which fonts looked best in which colors. By three and a half, he was using the computer to type out train schedules. Before he turned four, he was writing out paragraphs.

He looked for any opportunity to calculate. At first, he zeroed in on dates. William began asking people their age and month of birth, information he used to compute birth year. Around the time he turned four, he got his hands on a perpetual calendar and began calculating the day of the week on which different events occurred. When the family's Apple TV displayed the year in which a particular song came out, William would report how old his mom had been that year and how old he had been, a calculation that often drove him into negative digits.

By the time he was four, William could multiply two-digit numbers together. He could complete multiple-step problems in his head. At four years and three months, his parents videoed him swapping math problems with his uncle, beaming between bites of applesauce as he answered "5" when his uncle asked him to solve 9 squared minus 76. "I guess everybody's smart," William said afterward as he ran from the table. "Oh, not quite as smart as you, my goodness," Lucie said to herself, taking a deep breath as she wiped off the table and picked up an empty applesauce cup. On a standardized test he took a few months later, at four and a half, William scored better than 99.9 percent of five-year-olds who took the test.

For all of his impressive—more than impressive, sometimes shocking—intellectual abilities, William struggled to integrate in a classroom setting. He enrolled in nursery school with a therapist, just as Alex had done. He had mostly overcome his language delays by then: he had more typical back-and-forth exchanges with others; he initiated conversations and asked questions. But following a schedule was a feat beyond William's abilities. He resisted leaving behind any project he hadn't finished. If a teacher tried to help him put things back in place to move on to the next activity, he would swat her hand away.

In junior kindergarten, he hardly ever interacted with the other children. He rarely participated in group activities without prompting; he preferred to let his own interests dictate his pursuits. He had trouble remaining seated, struggled with a short attention span, and frequently disrupted other children's play.

He still couldn't take care of himself. He couldn't get himself dressed or undressed. While he had the dexterity to button his shirt and tie his shoes, he struggled to stay focused on the task long enough to complete it. When dressing to go outside for recess during senior kindergarten, he was frequently distracted; recess often ended before William had a chance to go outside. Even when his parents broke down what he was supposed to do at home—walk to your drawer, open it, get a T-shirt, and pull it over your head—William would wander off before he finished the task. "He's off in his brain," Lucie said. "It's such an amusing place that it's really, really hard to keep him in the here and now sometimes."

But over the next couple of years, things got easier. He began taking an ADHD medication, and his ability to stay on task improved. His doctor tinkered with his dosage and added an anti-anxiety medication to the mix. More improvement still. Eventually, William could make it through the school day without his therapist. He still needed more prompting from the teacher than most other kids to complete tasks, but it was progress.

Along the way, Lucie and Mike realized that William had an epic ability—and desire—to process novel information. He craved new topics. The richer in detail, the better. Lucie and Mike fought to keep a constant stream of dense reference books at the ready for William's birthday and Christmas gifts. Upon receiving a new one, he grew giddy. He got a glint in his eye and threw himself at the material.

Lucie and Mike saw that glee when William pored over the atlas he received for his fifth birthday. He was soon inhaling geography. He studied maps—maps of the world, maps of Canada, maps of Ontario. Then he discovered Street View on Google Maps and began spending long periods of time traveling around the world, street by street. When Lucie asked whether he'd like to visit those places in person one day, William told her no, he liked traveling from his living room.

William wasn't much of a showman—other people's approval meant relatively little to him—so he was generally content to sock information away in his brain for his own enjoyment. But every now and then Lucie and Mike caught a glimpse of what was bubbling beneath the surface. It happened one day when Lucie drove the boys out into the country-side to pick out a pumpkin, a month after William got his atlas.

"Why don't we just go to the grocery store?" Alex whined. "There's a box of pumpkins right by the door." Lucie pressed on. As they drove onto a rural road, far out of their usual circuit and into an unfamiliar part of Canada, Lucie jokingly asked the boys if they knew where they were. "Yes," William piped up. "We're on page 34." Lucie sat for a moment in silence. "Oh, wait, we just drove onto page 35." Lucie realized that William had done more than just study the atlas—he had *memorized* it. And not just their own neighbor-hood or their familiar haunts. He had memorized the entire thing.

Lucie couldn't probe too deeply; if she tried to pry open the doors to William's mind, he was quick to change the subject or shift the conversation back to something that interested him. But William's knowledge wasn't something he could conceal, either. His command of geography popped up again when Lucie once referred to a street in a place they were visiting. William supplied two cities that had a street of that name (in one of which the street was so short, few cab-bies would have known of it) and asked which one she meant.

William had the same fascination with vocabulary. The fall after he turned five, he saw Lucie pick a dictionary up off the shelf. William asked what it was. Lucie explained that it was a book that listed every word in the English language. "He looked at me like, 'How could you keep this from me all these years?'" Lucie recalled. When he received a dictionary of his own for Christmas, he abandoned all his other gifts and paged through it, soaking in all of the information. He discovered sections previously unknown to Lucie: lists of the most common words, lists of the most commonly used letters. "I would just go in and look up a word," Lucie said. "I never read it like a book."

It made him an ace at Scrabble—he knew every two-, three-, four-, and five-letter word that contained the letter q and delighted in slamming them down on triple-word-score spaces. He enjoyed creating words so much that sometimes he played both players in a two-player Scrabble game.

When William was five, Mike gave him a printout of the periodic table of elements. By the end of the day, he had memorized it. Lucie tried to coax William into playing outside, and she eventually persuaded him by giving him a piece of chalk he could use to draw on the ground. He took it and re-created the periodic table on their backyard patio. Over the next few months, he taught himself everything he could about each element: its symbol, molecular weight, electron shell structure, diameter in picometers, density, state of matter, year and country of discovery, what it is used in, percent on Earth's surface, and percent in our bodies.

His memory—at least for things that interested him—seemed to know no bounds. At five, he discovered pi. Over two days, he memorized it to the seventy-fifth decimal place. He memorized every country in the world, including information about its location, capital, population, population density, and area in square kilometers. He devoured a book on the flags of the world, memorizing the flag of each country, its year of creation, color scheme, historical changes to that color scheme, and the flag's size ratio.

His memory wasn't just photographic; it was episodic. It was as if he had a video recorder running all the time, capturing everything he saw, time-stamping his every activity. When a first-grade teacher asked William to write down a time he had gotten hurt, William recalled the precise date two years before when he had scraped his knee. One day, in the family's Lego room, William pointed to a Lego replica Alex had built of a character in one of William's books and said that Alex had built it a year ago that day.

He could slice and dice the information in almost any configuration. He knew how many elements were discovered in France. He could rattle off a list of fonts that began with any given letter. If you gave him a year, he would re-create the then-current version of the periodic table. He could tell you which of the world's flags featured any particular color.

William had always enjoyed music; he was a constant hummer, which irritated his brother. Around the time he turned six, he requested piano lessons, and his parents signed him up. His teacher, Kathy, reported to Lucie that William had perfect pitch. Over a thirty-eight-year career, she had seen a handful of students who supposedly had this same ability. These students could, perhaps, identify a note played in isolation without any benchmark. But she had never seen anyone like William. "We played this game all the time where he shut his eyes and I would play notes and chords and he would name them immediately," Kathy said. "I would try to trick him, but I never could."

Though he almost never practiced, he picked up new pieces quickly and could swap the right-hand part with the left-hand part, something Kathy had never seen anyone do. William constantly composed music, and he told Lucie that he had more than a hundred original songs on the CD in his head. He wouldn't perform the pieces on command, but he knew the precise length and title of each one. "There's no child like him, that's for sure," Kathy said.

He was still a kid. When he wasn't reading reference books, he loved stories about mischievous cats and neurotic squirrels. He laughed with his brother, stuck his tongue out at the table, got his face covered in chocolate, and delighted in potty humor; nothing cracked him up like *Calvin and Hobbes*. He dressed up like an alien for Halloween and donned a Santa hat around the holidays. When he worked in his "office" at home, a stuffed animal often tagged along. But anyone who knew William, anyone who'd gotten a glimpse at his mind, realized that the kid chasing bubbles in the elephant T-shirt wasn't your typical six-year-old.

⁀

Things were better. Things were oh-so-much better than they had been when Lucie was staying up at night wondering whether her boys would ever talk, laugh, or make a friend.

But she still had so many questions.

What yanked Alex out of the depths of autism? Maybe it was the jump start they had gotten on intervention. Maybe Portia housed the secrets to autism recovery. Maybe it was something internal to Alex, some part of his wiring that destined him to speak, to laugh, to connect, and nothing she or Portia did made any difference. Maybe it was only a matter of time.

William was no less a mystery. Was he actually autistic? Lucie—and some of his doctors—were uncertain. Maybe he was. There were plenty of symptoms you could point to. But maybe he was just extraordinarily bright and had a severe case of ADHD coupled with anxiety. There was really no way to be sure.

One of the psychologists William had seen was heartbroken for him. Sure, she was blown away by his academic abilities. But who could be his friend? Who could possibly keep up with him? At best, she thought he might find a peer group in graduate school. She told Lucie that her job was to get him there as fast as possible. But was that really the best she could do for her son?

The parenting books, of course, were still no help. Trying to hash out these questions with friends was difficult. The usual give-and-take of swapping parenting stories, offering and receiving advice from people navigating parallel developmental milestones, was almost impossible. As far as Lucie knew, no one else's kids were on this path.

That January, when Lucie watched a recording of a *60 Minutes* episode that had recently aired, she found someone who was. The episode featured a child prodigy, Jacob Barnett, who could've been William's twin.

Around the time Jacob turned two, he had stopped speaking; he stopped making eye contact, just as William had. He, too, was diagnosed with autism. Jacob and his parents had been through the same assembly line of therapists; they had endured the same devastation as their little boy disappeared into a world of his own.

There was the same turnaround. Language developed. Jacob began scouring books for new information, anything to keep his brain stimulated. He loved talking about physics but couldn't keep his shoes tied. He had a copy of the periodic table hanging in his bedroom and memorized digits of pi for fun. Lucie thought Jacob even *looked* like William.

About two-thirds of the way through the segment, the reporter introduced Joanne Ruthsatz, an assistant professor at Ohio State University who had been studying child prodigies for more than a decade. She talked about the prodigies' astounding memories; she believed there was a connection between prodigy and autism. Finally, someone Lucie could talk to.

After watching the rest of the piece, Lucie posted a link to the story on Facebook and then did a little Internet research. She tracked down Joanne's bio and a description of her work; then she sent Joanne a message telling her a bit about her children. She provided her contact information in case Joanne wanted to get in touch.

Within half an hour, Joanne dialed her number. Lucie described all that had happened to her boys. Joanne listened carefully. She asked questions and never doubted Lucie's story. She was convinced that William was a child prodigy. But what did that term really even mean?

That turns out to be a surprisingly difficult question to answer—even for the experts.

Chapter 2

What *Is* a Prodigy?

David Henry Feldman has gray hair, dark, round glasses, and a knack for storytelling—pause here for emphasis; home in on dramatic moments. For the last four decades, he's been a child development professor at Tufts University. His office sits on the outskirts of campus in a small, unassuming building in Medford, Massachusetts. There, Feldman pursues his studies with the zeal of the academic convert he became more than fifty years ago during a college class with a particularly charismatic professor.

Feldman has been a leading authority on child prodigies since the 1970s. He's also the first person to define what, exactly, a child prodigy is. As articulated by Feldman in 1979 (and slightly modified in the 1980s), a prodigy is a child who has reached professional status in a demanding field before the age of ten. For nearly forty years, academics have used this definition to gauge prodigiousness. How Feldman came up with it, though, is a story in itself.

Feldman has always been a bit of an academic maverick. He often ignored the details of assignments as a graduate student and the demands of department politics as a professor in favor of chewing on whatever academic puzzle consumed him at the time. When he was a young assistant professor at Yale, that puzzle had to do with the work of the Swiss psychologist Jean Piaget, a man made famous, in part, by his theory about the phases of childhood development. Piaget believed that all children marched through a number of developmental stages before eventually reaching the formal operational stage of development—a stage characterized by the ability to reason abstractly.

Feldman was convinced that Piaget had missed something. He was certain that there were exceptions to Piaget's theory. Surely child prodigies, for example, progressed through Piaget's stages of development more quickly in their fields of expertise than they did in other, unrelated areas. "So, I thought, okay, I'll go and read the literature on child prodigies, and I'll fill that in with a little empirical finding. I went and looked in the literature, and I could find nothing—virtually," Feldman recalled from his office. "It turned out that there hadn't been anything done by way of what we would call social science research, or any kind of science research, on child prodigies for more than fifty years."

Still, Feldman was determined to proceed with his project. He wanted to find some prodigies and give them a handful of psychological tests to prove his anti-Piaget point about child development.

But what *is* a child prodigy? The few academic papers Feldman had found didn't offer much guidance. Everyone seemed to recognize that there was something distinctive about these children, but early researchers resisted laying out precise criteria. Instead, these pioneers of the field took an "I know it when I see it" approach to identifying prodigies.

In one of the earliest of these studies, Géza Révész, a lecturer at the University of Budapest, wrote of his work with Erwin Nyiregyházi, a Budapest-born child piano player who gained local acclaim for his musical abilities in the early twentieth century. Erwin was exceptionally pale and slender with a long, thin nose and full lips. He tried to imitate singing before he was one and could reproduce melodies before his second birthday. By the time he turned three, he could play any song he heard on the piano. He began piano lessons and learned to read music at four. While he was still a child, his playing was heralded in Budapest, he sought training in Berlin, and he performed several times in Copenhagen and London for audiences that included the British prime minister and the royal family.

Révész viewed Erwin as an exceptionally rare breed, a child who defied nearly all categories. The one person to whom Révész thought Erwin *was* very similar was Mozart, perhaps the most famous of all child prodigies. Like Mozart, Erwin took to music at an early age, and his abilities developed rapidly. Both were of above-average intelligence and displayed a passion for music and remarkable creative ability. From Révész's descriptions, the reader gets the sense that Erwin and Mozart were something entirely different, a breed all their own, the contours of which science had not yet begun to define. It was perhaps for this reason that Révész referred to Erwin as a musical prodigy but skipped out on specifying what that term meant.

The next notable contribution to prodigy studies was a book-length examination of a child artist, a geographer, a dancer, a chess prodigy, and five music prodigies published in 1930. The study's author, Franziska Baumgarten, linked the term to talent and achievement, but just like Révész, she never formally defined it.

Leta Hollingworth, an early twentieth-century researcher, somewhat accidentally became another leader in the field. She didn't necessarily see herself as engaged in the study of child prodigies; her objective was to study high-IQ children. Some of her subjects certainly sound prodigious, though, and her work is generally included in the canon of child prodigy literature. Perhaps not surprisingly then, her research put a high-IQ spin on child prodigies, but she never actually defined the term (at least not in any scientific sense).

Hollingworth's journey with these children began during a psychology course on the mentally impaired that she taught at Columbia University's Teachers College in 1916. For the sake of contrast, Hollingworth decided to put an exceptionally bright child through a round of intelligence tests in front of her class. Two of her students nominated an eight-year-old boy, Edward Rochie Hardy Jr., who had an excellent academic record.

Hollingworth tested Edward using the Stanford revision of the

Binet-Simon intelligence scales. The original Binet-Simon version had been created to identify children of "subnormal intelligence." But over time, the test, which eventually became known as the Stanford-Binet, was revised to measure high as well as low levels of intelligence. In the modern version, scores fall along a smoothed-out bell curve. But the version that Hollingworth used compared the subject's "mental age" (a calculation based on the quality of the answers given) with the subject's actual age to calculate an intelligence quotient, or IQ.

Hollingworth began testing Edward in front of the approximately thirty students in her class. As soon as Edward answered the first question for his age-group, Hollingworth realized that she had gotten more than she bargained for. She immediately jumped forward to the questions for older children, which Edward also answered easily. It took two full class periods to test Edward, and even then he exhausted the scale without being fully measured, reflecting an IQ of at least 187.

As Hollingworth grew more familiar with Edward's background, she learned that he didn't speak until he turned two, but when he did begin to talk, he could say all the words he knew in German, French, Italian, and English. He was reading books like *Peter Rabbit* before he was three. By the time Edward was eight, he had picked up several more languages, worked out the Greek alphabet from an astronomical chart, and thought the ultimate in fun would be to have statistics for a country he had imagined on Venus.

Hollingworth began searching for other children with IQs above 180. Over the course of twenty-three years, she found only eleven others who hit her skyscraping IQ benchmark; she eventually deemed it "nearly useless to *look* for these children, because so few of them exist." According to Hollingworth's estimation, such children were fewer than one in a million.

Hollingworth described Edward as "prodigious." But when asked to clarify what she meant by that, she said that she merely meant that his abilities were "wonderful" or "extraordinary." For future prodigy scholars, it wasn't much to go on.

⟶

For Feldman's anti-Piaget purposes, selecting participants was critical. He needed something more specific than "wonderful" or "extraordinary."

Piaget thought that children didn't typically demonstrate the highest level of cognitive development—formal operational thought—until they were eleven. Feldman had his heart set on finding children doing just that in their field of expertise before Piaget thought it possible. So, in his initial prodigy study, he decided to include only children who performed at the level of a professional (which, he thought, clearly evidenced formal operational thought) before reaching double digits. In doing so, he laid the groundwork for what would become the leading definition of a child prodigy.

Feldman found three children who met his prodigy threshold—two eight-year-old chess aficionados and a nine-year-old composer. He gave them each four psychological tests that measured traits ranging from spatial reasoning to moral judgment. Then he declared victory: The prodigies' abilities with respect to these "general developmental regions" were in no way as exceptional as their skills in their areas of specialty. They were performing at the highest Piaget level in their area of expertise but thinking like their age-mates in all other ways. At the time, Feldman felt that he had proven his point (though he would later change his mind regarding whether the prodigies were actually demonstrating formal operational thought).

Feldman's prodigy definition served his purposes well. As the guiding threshold for an entire field of study, though, there were some bugs to work out. Child prodigies presumably have unique internal wiring that leads to their astounding abilities. Any study of child prodigies would ideally screen for children with such wiring and exclude others. But according to Feldman's definition, the ultimate yardstick for prodigiousness is a child's level of achievement—an imprecise indicator of the child's intrinsic abilities.

Still, Feldman's definition provided an anchor for a field that had long drifted through the scientific backwaters. With a definition in place, other researchers had somewhere to start. They could use Feldman's definition, critique it, or argue for alternative criteria; a more rigorous scientific debate could begin.

Feldman had meant to move on from child wonders once he finished his initial experiment. "I said it many times, that I don't care about child prodigies. I'm not particularly interested in them," Feldman recalled. "But that was just denying what was really true at another level."

He couldn't resist the inherent mystery of children with remarkably advanced abilities. He expanded his initial prodigy study to include three additional children: a writer, a mathematician, and a jack-of-all-trades prodigy who inhaled languages, math, science, and music. This last prodigy, Adam Konantovich (a pseudonym), didn't technically satisfy Feldman's prodigy criteria. Adam seemed preternaturally talented in many areas (one psychology professor said Adam was the most gifted child ever to take the Stanford-Binet) but didn't yet have a single area of specialty. Feldman reckoned with his definition of prodigiousness and decided that it was the sort of rule that could bend a little bit. As he put it in *Nature's Gambit,* the book he and his collaborator, Lynn T. Goldsmith, wrote about their prodigy research, "any theory worth its salt should be able to say something about talents like Adam's." So he included him in his study.

The difficulty of using such criteria to identify prodigies is further illustrated in a dissertation written by one of Feldman's students. The author, Martha Morelock, studied two children who she believed had IQs higher than 200—one of whom was a prodigy, and one of whom was (technically speaking) not.

Bethany Marshall (a pseudonym) qualified as a child prodigy. Bethany was born limp and not breathing. The doctors revived her, but she stopped breathing again later that night. The doctors discovered blood in her spinal fluid—the product, they said, of a blood vessel bursting under her skull during her rapid delivery. The doctors warned her parents that she might have suffered other damage.

Bethany began speaking at eight months, and when she was nine months old, she grew captivated by letters. She began memorizing verse at eighteen months and whole books from the library around the time she turned two. She wrote stories at three and poetry at six. The words seemed to come to her mind spontaneously; initially, they were accompanied by music. An adult writer judged the poetry Bethany had produced from the time she was eight equal to that of an adult professional—hence a prodigy, according to Feldman's definition.

Michael Kearney, on the other hand, technically missed the prodigy cutoff. Like Bethany, Michael had a traumatic birth. His birth weight was a mere four pounds, and he had an APGAR (the score used to assess a baby's condition after birth) of two out of a possible ten. His parents were told that he would be developmentally retarded.

Michael began talking at four months. At eight months, he told his parents when pears were on sale at the grocery store and requested that they buy Campbell's soup. His parents asked him not to speak while they were shopping because people looked at them strangely. Michael gave his mother driving directions before he turned two. He reported that he composed songs in his head, sometimes improvisations on entire symphonies, but he didn't know how to write the music down. Michael graduated from high school at six and college at ten. He went on to earn master's degrees in chemistry and computer science before enrolling in a chemistry Ph.D. program.

Bethany and Michael were both extraordinarily talented. Both endured traumatic birth experiences, evidenced a voracious hunger for knowledge, developed at breakneck speed, and spontaneously heard—and maybe even composed—music. But because Bethany had a specialty in which she reached professional status before the age of ten, she qualified as a prodigy, while Michael, who graduated from college at ten (not technically a professional), did not.

It's an imperfect definition. It's based on accomplishments (an external rather than an internal guidepost), and the age cutoff—which was eventually softened—is somewhat arbitrary. Feldman didn't even really mean for it to set the bar for prodigiousness—at least not initially. "I think it had some value," he said. "I didn't do it for the purpose of having it be a definition that would work for the field for all time. I did it because of what I was up to."

⟶

After Joanne met Garrett James and his autistic cousin Patrick, she graduated from Case Western and then hopscotched through a series of academic positions. Along the way, she wrote up her work with Garrett, but she kept quiet about any potential link between prodigy and autism. After all, her idea that the two were somehow connected was little more than a hunch.

That hunch was a highly unorthodox one. In 2005, when Joanne was on the cusp of initiating a small prodigy study, no one searching for the underpinnings of child prodigies' talents had suggested that they might have something to do with autism.

But Joanne was determined to test the waters. Over the years, she had contacted the families of talented children she read about online or in the newspapers. She spoke with the mother of a child who was memorizing books at fourteen months and began adding at eighteen months. She reached out to the parents of a little girl whose paintings were selling for tens of thousands of dollars before she hit double digits. They were fascinating cases, but were they prodigies?

Feldman's prodigy definition was a good starting point. But Joanne thought that any connection between prodigy and autism was likely genetic, so she wanted to use a standard that was more closely focused on factors intrinsic to the child and less oriented around a particular level of achievement.

The thing that seemed most critical, she thought—the dagger that cut to the heart of what it meant to be a prodigy—was the accelerated development of talent during childhood. That insatiable drive, which the psychologist Ellen Winner has described as a "rage to master," seemed less dependent on attributes of the child's field or parents; it seemed more likely to have a distinct biological engine behind it.

Such swift achievement often attracted a fair bit of limelight. A hodgepodge of people from journalists to scientists clamored to applaud children who skyrocketed to the top of a typically adult field. National or international acclaim by adolescence, Joanne reasoned, might serve as a proxy for the runaway-train development pattern she thought characterized prodigy. It was still a behavior-based definition, but by focusing less on whether the child had achieved professional-level success, she tried to create a threshold more closely aligned with attributes intrinsic to the child.

Interestingly, the imprecision of using behavioral criteria to make a diagnosis is something autism researchers know all too well. Autism, like prodigy, is diagnosed based on external, observable symptoms, not genetic tests, blood samples, or brain scans.

Austism was described in the 1940s by two researchers, Leo Kanner and Hans Asperger, working on two different continents. Kanner was a psychiatrist who founded the Johns Hopkins Children's Psychiatric Clinic, the first such clinic in the country. In 1938, Kanner met Donald T., a five-year-old from Mississippi. Donald paid no attention to the people around him, and his mother reported him happiest when left alone. Objects, on the other hand, fascinated him. Donald would spin anything and jump in excitement as he watched it rotate. His days were packed with ritual and repetition. When he spun a block, he started with the same side facing upward; he always threaded buttons in the same sequence. He had an excellent memory for faces and names, and he memorized a large number of pictures from the encyclopedia. His speech consisted almost entirely of parroting phrases he had heard before; his words often seemed unrelated to what was going on around him.

Over the next few years, Kanner saw several other children exhibiting the same "extreme autistic aloneness." Such children were often labeled feebleminded, idiotic, or schizophrenic, but Kanner was convinced that what he was seeing was a unique condition. He published a paper reporting on this new condition in 1943; the next year, he named the syndrome early infantile autism.

Hans Asperger had the same idea at almost the exact same time. While working at a children's clinic in Vienna, he encountered a number of children who he believed shared a syndrome marked by social withdrawal and communication difficulties, such as problems with eye contact or an unusual speech pattern, as well as a high degree of creativity. Asperger emphasized that this syndrome could affect individuals of all levels of intellect, including the "highly original genius." He began using the term "autism" to describe what he had seen as early as the 1930s, and in 1944 he published a paper arguing that the collection of behaviors he had observed constituted a new, independent condition. He called it autistic psychopathy.

The conventional wisdom is that Kanner and Asperger were working independently and that their nearly simultaneous identification of autism—and the nearly identical names they gave the syndrome—was a grand coincidence of history. This view is shifting, however, in light of new evidence that Kanner worked closely with former colleagues of Asperger's and thus might have been more familiar with Asperger's work than he let on.

But regardless of who identified and named autism first (and who knew what when), both Kanner and Asperger placed behavioral abnormalities at autism's core. Kanner emphasized that the children's "fundamental disorder" was the "inability to relate themselves in the ordinary way to people and situations from the beginning of life." Asperger thought a key marker of the condition was "the shutting-off of relations between self and the outside world."

A diagnosis based solely on behavioral symptoms is inevitably slippery (just as Feldman discovered while selecting participants for his prodigy research). In the early autism studies, scientists used differing criteria for autism. Two decades after Kanner first identified autism, he complained that it had become "a pseudodiagnostic wastebasket for a variety of unrelated conditions." And this issue continues to plague autism, as demonstrated by the ever evolving *DSM* diagnostic criteria: autism has gone from a symptom of schizophrenia to a condition independent of schizophrenia; the diagnostic criteria have shifted over time such that they encompass an expanding number of people.

When work began on the most recent edition of the *DSM*, there was talk of developing diagnoses based on the underlying neuroscience rather than symptoms. But the new *DSM* offered no drastic rethinking. The drafters slotted some diagnoses into different chapters, combined some separate diagnoses into single conditions, added new diagnoses, eliminated others, and refined diagnostic criteria. But fundamentally, the *DSM-5* maintained its symptoms-based approach to diagnosis rather than focusing on the underlying biochemical mechanisms.

The entry for "Autism Spectrum Disorder" is a combination of what had been four distinct diagnoses (autistic disorder, Asperger's disorder, childhood disintegrative disorder, and pervasive developmental disorder not otherwise specified) in the previous version. According to a fact sheet published by the American Psychiatric Association (the organization that publishes the *DSM*), this change was made in part because researchers didn't think that these diagnoses were applied consistently across clinics and in part because it was thought that a single diagnosis better reflected the idea that autism symptoms vary in severity. But neither of these reasons has much to do with whether these symptoms actually stem from one distinct underlying abnormality or four.

This emphasis on behavior leaves scientists in a quandary. As the prominent autism researcher Uta Frith once observed, "Behaviour, however reliably it is measured, is not revealing about its cause. There is a mapping problem. Many different causes can underlie the same behaviour. On the other hand, behaviour that looks different in different individuals may actually be due to one and the same cause."

⟶

Beyond the challenge of pinning down the definitions of prodigy and autism, Joanne faced another hurdle: her research pockets were empty, so she couldn't travel to the children and their families. She had tried to get NIH funding, but her application had been rejected.

She still wanted to do *something*, though, to at least find out whether her theory had merit. She didn't need ironclad proof. She just needed something quick and dirty that could be done on the cheap to explore whether there was anything to the idea that autism and prodigy were connected.

The answer came from abroad. The Autism Research Centre, an organization nestled within the University of Cambridge's psychiatry department, operates as a hub for scientists studying autism. At its core was Simon Baron-Cohen, one of the most famous and prolific autism researchers in the world. In 2001, he and his colleagues had published the Autism-Spectrum Quotient, or AQ, a test that measured a handful of traits associated with autism.

It was a snappy little test. With just fifty questions, it was quick and to the point, and it was designed to be self-administered. Most important, it was free.

Joanne thought she could use it as a canary in the mine for her theory. Autists' family members often demonstrate autistic traits, just not at a level that would merit an actual autism diagnosis. Joanne reasoned that if there was a connection between prodigy and autism, maybe the AQ would show that the prodigies' relatives, too, had some autistic traits, just like the autists' relatives. She decided that if they did, that would be her sign; she would launch a full investigation into the connection between prodigy and autism.

An autism researcher Joanne had known since high school agreed to make the AQ available to her patients and their families. For a control group, a local school distributed the AQ to families whose children had no documented disabilities. Joanne mailed the AQ to the families of some of the prodigies—those children who showed the accelerated development she was looking for—with whom she had been in contact over the previous few years. She wound up receiving ten completed surveys from each group.

The responses—though relatively few—were intriguing. The autists' relatives and the prodigies' relatives' AQ results indicated that both groups had elevated levels of some autism-linked traits. Both had more difficulty with "attention switching" (the ability to multitask, switch between activities easily, and embrace spontaneity) and were less drawn to social situations than the control families.

But the headline result, the one that Joanne thought might offer a real clue to the relationship between autism and prodigy, came from the families' scores in attention to detail, another autism-linked trait. The relevant AQ questions measured the degree to which the test takers noticed small changes and absorbed patterns, dates, or other tidbits of information. In this category, the prodigies' families spiked. They demonstrated significantly greater attention to detail than the control families *and* scored higher than the autists' relatives.

The results reassured Joanne that she was onto something. The connection between prodigy and autism, the one she had suspected since seeing Garrett and his cousin Patrick together seven years before, was not a figment of her imagination. But there was also a fascinating twist. The prodigies' families' and the autists' families' results were not identical. The prodigies' relatives seemed to have less trouble with conversation and social niceties than the autists' relatives, at least according to the AQ. It was a curious inconsistency. It seemed that prodigy and autism were connected in some—but not all—respects. A summary of Joanne's findings was published in the fall of 2007 in the academic journal *Behavior Genetics*, revealing the first hints of a connection between prodigy and autism.

In 2008, Joanne began a tenure-track position at Ohio State's

Mansfield campus, where she finally got her hands on some research funding: she received a seed grant—more than $13,000—earmarked to help new faculty jump-start their research. With money in her pocket, she could finally travel to the prodigies and begin digging into their cognitive profiles. Once she did, the deeper connections between prodigy and autism quickly began to reveal themselves. The first of these was memory.

Chapter 3

The Tiniest Chef

By the time Greg Grossman was a toddler, the skinny New Yorker with the fluffy dark hair already had a wide palate. Fellow diners around Manhattan and East Hampton gaped, astonished, as the doe-eyed kid requested foie gras and other adult fare. "He'd order anything—anything different and weird," his mother, Terre Grossman, recalled. While waiting for his food, he scavenged for ways to observe the back-of-the-house action. He climbed onto his chair for a better view of the pizza oven; he snuck into the kitchen to watch the chefs at work.

Greg soon began experimenting in his own kitchen. At age four, he presented special candlelight dinners to his parents—often little more than a stuffed baked potato. At six, he used canned tomatoes to whip up his own pasta sauce. His recipes grew ever more sophisticated as he began sautéing fresh basil to add to his sauce and switched from limp spaghetti to al dente pasta. He began directing his mother as to what produce was in season when they went to the supermarket. "I would be buying stuff, and he would actually be telling me not to buy certain things, you know, 'This isn't ripe; it's not in season,'" Terre said.

Around the time Greg was nine, the complexity of his creations swelled. He began using a grill, an appliance he described in a school essay as an "infrared, propane masterpiece of stainless-steel." His repertoire exploded. He went from riffing on classic pasta dishes to pairing melon carpaccio with anchovies and wrapping sushi with foie gras.

That year, he prepared the meal that convinced him he wasn't just messing around in the kitchen. With his father out of town, Greg urged Terre to watch television while he cooked dinner for the two of them. Terre stuck her head into the kitchen occasionally to monitor her child's use of the stove and knives, but her worry was unfounded. Greg emerged from the kitchen unscathed, having crafted a meal that he would later declare his "first work of cooking art": pan-seared scallops with a balsamic vinegar glaze and a wild mushroom medley. "It was so unbelievably good," Terre said. "I totally freaked out."

After that, food was everywhere. The family television pulsed with culinary programs as Greg discovered a stable of talented chefs he admired: Jacques Pépin and Rachael Ray for bringing cooking to a broad audience; Andrew Zimmern for his use of exotic ingredients (Greg declared he would eat many of the bugs Zimmern featured but not the beating snake heart). Greg monopolized the computer, researching chefs and cooking techniques. He satisfied middle school world civilization assignments with reports on ancient food cultivation and completed science projects by experimenting with Scoville heat units and inspecting bacteria growth in kitchens. He saved his money to buy specialty food products, requested only professional chef products as gifts, and dreamed of owning his own truffle.

Kitchen implements became contested territory in the Grossman household. For months, Greg hounded his parents for a new knife set. At first, the Grossmans refused. When Greg received the coveted knife set from a family friend, Terre nearly returned it. But after hours of watching Greg work with the knives, his parents finally relented: he was skilled and invariably careful. Their next battles were over fire (Greg insisted he needed it for crème brûlée) and liquid nitrogen (essential, Greg said, for ice cream and cotton candy). Greg won them both. "I was a little scared. What if I'm not home and he's playing with some friends and something happens, the thing blows up or he burns somebody?" Terre recalled. "It was very, very difficult."

Hands-on training proved hard to come by. Most cooking classes that would admit a young person were geared toward beginner fare like introductory cupcake making. Greg eventually enrolled in a course for teens. His skill at the stove drew the attention of the instructors, who asked him to work at an upcoming benefit with them. Greg found an adult class that sparked his interest, but he couldn't participate because wine was served during the course. He would have to find a different way to learn.

Around the time he was twelve, Greg got a job busing tables and washing dishes at an Italian restaurant in East Hampton. When one of the prep cooks bailed on his shift, Greg stepped in, cutting carrots and peeling potatoes. He proved adept at the work, and soon he was filling in on this shift every week, eventually working as a line cook. When he was called on to run the pizza station, he made, by his mother's count, forty-two pizzas in one night without burning a single one. Fearing he would be teased, he kept his job a secret from his friends.

But it was the cafeteria at Greg's East Hampton school that really jump-started his education. The Ross School Café, an eatery committed to "regional, organic, seasonal, and sustainable purchasing and culinary practices," wasn't a meat-loaf-and-mashed-potatoes kind of place. The chefs used vegetables culled from the school's garden to produce spinach and shiitake mushroom salads and poached asparagus with miso scallion vinaigrette. For Greg, it was a haven. He lit out for the café every chance he got.

When Greg was in fifth grade, one of the chefs at the café lent him a copy of a book by Ferran Adrià. For decades, Adrià had helmed elBulli, a restaurant in Spain frequently heralded as the best in the world until closing its doors in 2011. The book included stunning photographs of foods prepared and presented in fantastically unexpected ways. Greg was captivated by it. The ten-year-old carefully wrapped the book in paper to protect it and then devoured its contents. He searched the Internet for an inexpensive copy so that he could have one of his own. "That's one of the things that really sparked my deep dive into cooking," Greg recalled. "That set me off on this quest for creativity and technique and trying to figure out what the book was about."

Greg began buying food equipment and chemicals and experimenting with cooking techniques unfamiliar to many professional chefs. He hunted down information related to the science of food preparation and dragged his mother to food trade shows in pursuit of liquid nitrogen. He acquired cooking implements his parents had never heard of.

The following summer, having just turned a whopping thirteen years old, Greg announced that he would no longer attend sleepaway camp in the Adirondacks. The camp lacked air-conditioning, there were bugs everywhere, and, most damningly, the food was terrible. One day in June, Greg accompanied his mother on a trip to Vered Gallery, an upscale art gallery in East Hampton where she had a business meeting. Greg perked up at the mention of an upcoming event at the gallery, a silent art auction fund-raiser.

You're having an event? he asked. Who's catering? We just cut up cheese and do a few little things, the gallery owner told him. There is no caterer. Give me a credit card and a budget, Greg offered, and I'll get together a few servers and do four hors d'oeuvres. The co-owner gave Greg $100 to work with. Just keep it semi-kosher, she told him.

Greg hired two girls from school and told them to wear black dresses and black shoes. He hired a sous chef, a friend he was working with a lot at the Ross School Café.

On the night of the event, Greg told his classmates turned servers how to pronounce the hors d'oeuvres and explain the ingredients. Relying on a tidbit culled from the Food Network, Greg told them that they needed to know what they were serving to avoid any allergy mishaps.

For his first catering job, Greg served salmon gravlax with shiso crème fraîche—a raw, cured salmon appetizer. Other offerings included Thai chicken satay with citrus turmeric tzatziki and Indian spiced hummus with black sesame and curried nacho chip. For dessert, he prepared fresh whipped cream with peach gelée and mixed berries in a phyllo cup and chocolate mousse with lychee foam in a phyllo cup.

After the event, the co-owner complimented Greg on his execution. The event was beautiful, she told him. Everything was wonderful. Her only complaint was that the citrus turmeric tzatziki sauce he served with the Thai chicken satay wasn't kosher—an infraction that became a running joke because Greg is Jewish.

For the rest of the summer, Greg hawked his services around the Hamptons, making pitches to prepare food for parties and openings. When someone expressed interest, Greg proposed recipes more or less on the fly. He cooked a fiftieth anniversary dinner for thirty people and handled the food preparation for a friend's grandmother's party. Vered Gallery hired him back for another exhibition in August, and Greg prepared truffles and fish soufflés. By the end of the summer, Greg estimated for the *New York Post* that he had prepared seventy-five pounds of scallops, thirty pounds of salmon, and two hundred micro-green salads.

Greg had an eye for the business side of things. He made his own contracts and insisted on photographer credits for pictures he provided. He found a nondisclosure agreement online, sent it to a friend's lawyer father for edits, and began using it before he would discuss industry-related ideas with potential partners. He did his own billing, created budgets, and sought out wholesale suppliers to help keep costs down.

He plunged into the circuit of food industry events. At the James Beard Foundation's Chefs and Champagne event, a glitzy tasting party in the Hamptons, Greg hobnobbed with fellow chefs, dove into conversations about the nitty-gritty of ingredients and food prep, and gave a cooking demonstration and interview for a TV station covering the event. At the International Restaurant and Foodservice Show in Manhattan, he introduced himself to vendors and suppliers, listened to the industry legend Danny Meyer speak, and demoed a Pacojet, a machine that micro-purees frozen foods. He buzzed his way through the James Beard Foundation Awards at Lincoln Center, attended the New York International Gift Fair, and participated in the StarChefs convention.

He made the biggest splash at the National Restaurant Association Show, a Chicago event that NBC declared the thirteen-year-old Greg's "coming out party." Greg demonstrated how to whip up a smoked Maine lobster tail with orange/anise granita, carpaccio of melons, and whipped shiso oil. Taking a page from his favorite TV chefs, the eighth grader included instructions on simpler methods for preparing the same dish at home. His public relations team beamed from the sidelines. Rumblings of a possible TV deal filled the air.

Greg rode a media swell back to New York. The newspapers ribbed him for being younger than a 1994 Bordeaux and dissected his past experience. When TV came calling, Greg's teen idol looks didn't hurt. His hair—dark brown, thick, and wavy—garnered Justin Bieber comparisons. His eyelashes—long and dark—"would make a Jonas brother jealous." Greg cooked on the *Today* show, appeared as a guest on Fox News Channel's *Your World with Neil Cavuto,* and chatted with Gayle King on Oprah Radio.

Just one summer after his Hamptons catering debut, the Greg Grossman cooking cyclone reached dizzying speeds. In June, Greg whipped up 450 desserts—an innovative take on strawberry shortcake—for a school benefit. In July, he stopped in at an event to celebrate Flatiron chefs in Madison Square Park. Days later, he zipped off to Ohio to help raise money for an event at Veggie U, an organization dedicated to teaching kids about healthy foods, where he and his team churned out 650 tastings in an hour. After hustling back to New York, Greg helped fete a new coffee product by using liquid nitrogen and spray-dried yogurt powder to craft a dessert that looked like a frozen cappuccino. Next it was off to the French Culinary Institute to watch another chef demo before ordering specialty food products from out of state. He flew to Los Angeles to help with another food product promotion. After he returned to the East Coast, he ventured out to the Culinary Institute of America for an event with the Avant-Garde Cuisine Society in Hyde Park, New York, and then darted back to Manhattan for an industry gift show.

He was almost always in a white chef's coat emblazoned with the logo of the Culinaria Group—the organization he founded as a beachhead for his professional and charitable work. He still hadn't started high school.

—

By the summer of 2009, Greg Grossman was all over the newspapers, all over the Internet, all over television. It was only a matter of time before Joanne found him.

But was he a prodigy? He more or less met Feldman's standard: Greg performed at a professional level at a very young age. Joanne's slightly modified definition was satisfied, too; there was no denying Greg's accelerated development. The quality of his work and the depth of his knowledge were recognized by professional chefs at the top of the game.

But there was a question about his field of expertise: Can you be a *cooking* prodigy? Was cooking really as difficult to master as, say, theoretical physics or music composition?

Two words kept Joanne from forgetting the kid chef: "molecular gastronomy," a term sometimes used to describe Greg's style of cooking. It was no accident that it sounded scientific. The phrase was coined in the early 1990s to jazz up a conference on food science in Sicily. The conference's host wanted something that sounded weightier than "Science and Gastronomy," so the "International Workshop on Molecular and Physical Gastronomy" was born.

In practice, molecular gastronomy looks a lot like science. Hervé This, one of molecular gastronomy's chief practitioners, has created charts illustrating when coffee chills depending on when milk is added, spent more than three months researching the texture of egg whites used in soufflés, used nuclear magnetic resonance to analyze carrot-based soup stocks, and puzzled out how to uncook an egg (he said the key was to add sodium borohydride to detach the protein molecules from one another).

Such experiments aren't restricted to the laboratory. The restaurant luminary Pierre Gagnaire regularly incorporates This's ideas into recipes. Heston Blumenthal, the maestro behind the Fat Duck, a three-Michelin-star restaurant in Bray, England, explored the science of cooking on *Kitchen Chemistry with Heston Blumenthal,* a series of six half-hour programs. Exotic-sounding ingredients and techniques like liquid nitrogen (causes rapid freezing), hydrocolloids (substances that form a gel when mixed with water), and dehydration (removing the water from food) are the tools of chefs as statured as Thomas Keller of the French Laundry and Per Se, Grant Achatz of Alinea, and Greg's culinary idol, Ferran Adrià. Molecular gastronomy was all the rage. The chef as scientist was king.

Greg hated the term. He thought it implied flashy cooking. He preferred to call his cooking modern cuisine. The core idea was to enhance the flavor of the food, not to show off fancy ingredients for their own sake.

But molecular gastronomy or not, Greg's method of cooking was, at heart, a science experiment. He used the precision of a chemist to test temperatures, cooking times, and preparation methods, always trying to extract the best of an ingredient's flavor and texture. Greg wasn't slaving over equations in the back rooms of academia, but in practice his cooking was just a different expression of the same thing. Joanne decided that Greg Grossman was a prodigy. She wanted the kid in the kitchen, if she could get him.

Joanne clicked her way through articles on Greg and stumbled upon a story detailing his involvement with Veggie U. Greg had recently been in Ohio—a mere fifteen minutes from Joanne's house. She had just missed him.

She called her husband, Jim. He contacted a family friend who worked at the Chef's Garden, an organization connected with Veggie U, and got the Grossmans' contact information. A male voice answered Joanne's call; she asked if she was speaking with Mr. Grossman. When the caller confirmed, Joanne launched into her pitch.

There was something a bit uncomfortable about asking parents if she could study their child, but Greg's father immediately put her at ease. Her work sounded fascinating. He was interested. But about twenty minutes into the call, Joanne realized that the Mr. Grossman she was speaking to wasn't Greg's father; it was Greg. A fourteen-year-old with the voice and confidence of an adult.

Greg tried to reassure Joanne that it was no big deal—he did all kinds of cooking deals without his parents' knowledge—but Joanne hung up. Her Institutional Review Board, the organization that oversees research ethics, forbade her to talk to a minor about her research without the consent of his guardian.

Joanne waited until Saturday morning and tried the Grossmans again. This time, Greg's mother, Terre, answered her call. Joanne explained that she had inadvertently asked Terre's young son if she could put him through a battery of psychological tests.

Terre just laughed. She said it happened all the time. A major broadcasting station talked to him for three weeks before she even knew about the conversation. Joanne offered to travel to New York, but Terre told her there was no need: they would be back in Ohio for another Veggie U event the following July.

When Terre and Greg arrived, Joanne dropped Greg off at the Chef's Garden to prepare for the next day's competition and drove Terre back to her house. Over the next two days, Terre detailed Greg's development. The media was right: Greg was an astounding kid. But that was only part of the story. From Terre's perspective, the road hadn't always been smooth.

When Greg was born, Terre was forty-one years old, and her husband, Ed, was fifty-two. At six months pregnant, Terre had been bitten by a Lyme-infected tick in the Hamptons and panicked that her baby might be stillborn. A doctor put her on antibiotics, and the pregnancy continued.

Greg entered the world seven weeks before his due date. He weighed a mere four pounds, eleven ounces, but he was otherwise healthy. He began speaking at the usual time, but he garbled his words well into preschool, spitting out syllables that sounded nonsensical to all but his mother. Greg eventually corrected his speech through therapy and tongue exercises. School presented other problems. In early grade school, Greg often finished assignments quickly, leaving him with free time. He filled it by talking or joking around, and this often landed him in trouble with his teachers.

But from Greg's youngest days, an irrepressible industrious streak propelled him to learn, to do, to create. When Greg's parents bought him a battery-operated toy piano to play with during car rides, he quickly began composing little ditties, such as "Black Notes," a tune played exclusively on the sharp and flat keys of the piano. He demonstrated a knack for computers and, at four, declared himself the "Komputer Kid." He generated business cards to advertise his services and offered his computer expertise to friends' parents. A couple of years after that, Greg grew fascinated with clusters of rocks, shapes in the snow, and other natural formations that looked like human faces. He began photographing these as part of a venture he dubbed Naturefaces and developed plans to display them on a Web site, a calendar, and a movie. Eventually, Greg became fully engrossed with food, and his other interests fell by the wayside.

Greg's insatiable fascination with food unexpectedly provided common ground with his classmates. Greg loved to teach other kids the ins and outs of cooking, and Terre often found him and his buddies in her kitchen, making tortillas or fanning sushi rice. When a colleague asked Greg to store his anti-griddle (a cooking appliance with a surface that plummeted to -30°F), Greg invited classmates over to flash freeze their favorite foods, creating frozen chocolate pudding and olive oil treats.

The Grossmans sold their East Hampton home during the financial crisis and relocated to Manhattan. Greg finished the school year in the Hamptons, living on a friend's couch. But his need to cook didn't flag. Through all the upheaval—and his first year of high school—Greg grew ever more consumed by food. At school, he assembled a proposal for a culinary-focused course of independent study. After his idea was approved, he vacuum-sealed meat and then cooked it in a vat of water, flash pickled foods, and experimented with creating new textures using hydrocolloids. On his own time, he attended the International Chefs Congress, yukked it up with fellow chefs online, devoured forty-plus meatballs at the New York City Wine & Food Festival, and bemoaned the closing of *Gourmet* magazine. He launched the Amaya Project, the aim of which was to integrate food with other forms of art, with a ten-course meal; celebrated the end of the Food Network's banishment from Cablevision; and attended a competition at the Culinary Institute of America. It never occurred to Greg to take a break.

This constant call to create followed Greg when he traveled to Ohio for the Veggie U Food and Wine Celebration. It was a behemoth of an event. More than thirty chefs, together with their teams, participated. Greg and his team prepared thirty pounds of beef tongue as part of a dish that also included chickpeas, caramelized fennel, and a sugar snap pea broth. When the event ended, Greg got dropped off at Joanne's house. Like a moth to a flame, he went straight to Joanne's kitchen. He got a knife in his hand and vegetables on the table. Within moments, he was chopping.

⟶

Another piece of the prodigy puzzle fell into place as soon as Greg completed the Stanford-Binet IQ test.

He did well across the board, never dipping below the ninetieth percentile in any subtest and often hovering far above it. But it was the working memory result that again caught Joanne's attention. It's a score meant to reflect an individual's ability to manipulate (rather than merely recall) information stored in short-term memory. Repeating a series of numbers back to the examiner would measure recall, for example, while adding together the first three numbers in the list would measure working memory. On this subtest, Greg reached the tip-top of the scale, registering a score at the 99.9th percentile—just as Garrett James had before him.

And just like Garrett James, Greg had plenty of amazing-memory anecdotes. As a small child, he could listen to a complicated musical piece and then reproduce it from memory. For school plays, he memorized not just his own part but the entire dialogue; he could always be counted on to help classmates struggling with their lines.

When it came to food, he almost never forgot. His mind was awash with all he had learned about restaurants, chefs, and supplies, techniques he picked up on TV, and knowledge gleaned from food Web sites. His memory for cooking formulas, ratios, and recipes was sharp. When Joanne asked him to write down the recipe for one of the dishes he prepared at her house, Greg seemed surprised by the request: you'll remember it, he said.

It's well documented that experts have exceptional memories for information relevant to their specialty. In a groundbreaking 1946 doctoral dissertation, the Dutch psychologist Adriaan de Groot reported that expert chess players could recall the configuration of pieces on a chessboard with much greater accuracy than less skilled players. Since then, dozens of studies have examined the power of expert memory, and the same supremacy of experts for recalling facts relevant to their domain has been found in fields as varied as engineering and figure skating. The pattern holds for food and music, Greg's and Garrett's specialties: waiters demonstrate better memories for food and drink than nonwaiters, and musicians demonstrate better memories for music notation than nonmusicians. It was fairly predictable, then, that Greg would have a great memory where food was concerned and that the same would be true for Garrett with respect to music.

But it's equally well established that the memories of these experts tend to be notable *only* for facts relevant to their domain. Master chess players demonstrate superior recall for configurations of pieces that could emerge in real games. But when chess pieces are positioned randomly across the board, their recall is no better than that of weaker players. Psychologists have thus theorized that the memory advantage is due to the experts' greater experience and familiarity with their subject matter, *not* superior overall memory capacity.

What was interesting about Garrett and Greg, though, was that they weren't following the typical pattern for adult experts. Their memories were certainly finely tuned with respect to information relevant to their specific fields. But when Joanne administered the working memory section of the Stanford-Binet, she wasn't quizzing Garrett on music patterns or grilling Greg about recipes. She was reading off sentences or pointing to strings of numbers and listening as Garrett and then, later, Greg flawlessly repeated them back to her. The information had nothing to do with music, nothing to do with cooking. But the prodigies were unstoppable. Unlike the adult experts, the prodigies' working memories were excellent *in general*.

As Joanne moved forward with her research, the pattern she saw with Garrett and Greg repeated itself. Again and again, the prodigies earned exceptional scores on the working memory portion of the Stanford-Binet. It wasn't that they remembered everything. Many of the prodigies reported that their memories were nothing special when it came, for example, to names or faces or movie plots. But when they paid attention to a particular task, as they did during their IQ testing, their working memories dazzled.

This pattern seemed to suggest that the prodigies' abilities were somehow different from those of typical experts. But if the prodigies weren't operating like miniature adults, how to explain their abilities? Could the prodigies' extreme memories have something to do with the link to autism that Joanne was pursuing?

⌐

Hidden in the depths of the *DSM-IV,* within the pages devoted to autism, a mention of extraordinary memory could be found by the careful reader.

It would be easy to miss. Extraordinary memory wasn't listed in autism's two-page "Diagnostic Features" section. Nor was it mentioned in the page-long "Associated Features and Disorders" section. In a section labeled "Specific Age and Gender Features," buried in a sea of information—sandwiched between a description of social shortcomings and the higher incidence of autism among men—was a single mention of notable memory: "In older individuals, tasks involving long-term memory (e.g., train timetables, historical dates, chemical formulas, or recall of the exact words of songs heard years before) may be excellent."

Having briefly mentioned extraordinary memory, the *DSM-IV* quickly dismissed it. Even when an autist demonstrates outstanding recall, the manual said, "the information tends to be repeated over and over again, regardless of the appropriateness of the information to the social context."

From reading the *DSM-IV,* you might think that exceptional memory was hardly worth mentioning (and even this brief description was struck from the *DSM-5*). But the idea that some individuals with autism might display extraordinary memory can be traced all the way back to Leo Kanner and Hans Asperger, the two men credited with identifying the condition in the 1930s and 1940s. Kanner, for example, noted that many of the children he saw could recite "an inordinate number of nursery rhymes, prayers, lists of animals, the roster of presidents, the alphabet forward and backward, even foreign-language (French) lullabies." Asperger similarly observed that one of his subjects had an excellent memory for digits and that, among autists, there were some who could name the saint for every day of the year, young children who knew all the Vienna tramlines, and some who demonstrated "other feats of rote memory."

Similar reports of autists with extraordinary memories appear in popular reports and academic papers—a boy who memorizes movie release dates, another who memorizes train schedules. But systematic studies have revealed that memory in autism is complicated: autists' performances on memory tests vary across many dimensions, including the type of memory test, the nature of the stimuli presented, and the context in which those stimuli are presented. Extreme memory for at least some types of information seems to be a trait of some but not necessarily all autists. The memory of savants is an altogether different beast.

Nadia was born in Nottingham, England, in 1967, the second child of two Ukrainian immigrants. She said a few words before she turned one but then stopped speaking. She was sluggish and clumsy; she struggled to feed herself. She was extremely particular about her clothes and arranged her dolls and stuffed animals in a precise order on her bed. She was prone to violent tantrums, some of which lasted for two to three hours. Eventually, she was diagnosed with autism.

When Nadia was three and a half, her mother spent a few months in the hospital. Nadia was ecstatic upon her return. Without warning, she began drawing on the walls.

After that, she drew often. She drew quickly, dashing off lines and often finishing her drawings within only a few minutes. She would rip through several sheets of paper during a sketching session. She never checked her drawings against any sort of reference material; she relied only on her memory. At the peak of her obsession, Nadia drew everywhere: blank paper, lined paper, newspaper, picture books, cereal packets, and the tablecloth.

Her drawings didn't look like those of other children. A typical drawing of a horse by a six-year-old portrays the animal from the side; the image is static and simplistic. The horse may be distorted, its body stretched out, or it might resemble a table, a square body with four legs popping out from it.

But some of Nadia's earliest sketches portray a horse head-on. Her lines, almost always drawn with pen, capture the wild complexity of the horse's mane and depict some of the musculature of the leg. These pictures of horses—one of her favorite subjects—improved rapidly; she captured the animal at unusual angles and always depicted it with a sense of perspective. The frenzy of her lines captures the horses in motion: the animals appear arrested in mid-stride, ripped from the hunt, frozen while ambling along with a rider in tow.

When Nadia's mother first showed the child's drawings to a team of psychologists, they thought it was a hoax. Such drawings could not possibly come from six-year-old hands—especially not the hands of a child who was mute, tantrum prone, and otherwise uncoordinated.

Nadia was certainly unique in this way, but she was not alone. She was a savant, an individual with what Darold Treffert, a Wisconsin psychiatrist who has spent more than fifty years studying savant syndrome, has termed an "island of genius"—a spike in aptitude combined with a more general impairment. Sometimes this aptitude or talent is merely surprising in light of the individual's disability. Sometimes the savant's level of talent would be amazing even without the disability, as was the case with Nadia's drawing. It's the underlying disability, though, that technically distinguishes the savants from the prodigies; savants have one, while prodigies do not.

Savants display a varied collection of talents. One common specialty is calendar calculating, the ability to quickly and accurately determine the day of the week on which a particular date will fall. A famous pair of twin savants could perform this calculation forty thousand years into the future or the past, easily able to determine the day of the week on which July 23, 12,213, will fall. Other savants are particularly gifted at music, art, performing complex calculations, or building models and working with machinery. Leslie Lemke, for example, can perfectly replicate complicated music pieces after hearing them only once, despite being blind and having an extremely low IQ. George Widener, who was diagnosed with Asperger's disorder in his thirties, creates intricate artwork into which he incorporates dates and historical facts.

Over time, scientists realized that in many—perhaps most—cases, the savants' underlying disorder is autism. Treffert estimated based on his most recent study that *70 to 75 percent* of savants have autism. With figures like that, it's not a strange coincidence that Nadia, a child with developmental abnormalities and extreme drawing ability, was autistic; it's highly probable.

Savants also seem to have nearly infallible memories. This ability is so predominant among savants that Treffert has declared "massive memory" present in every individual with savant skills. An early, large-scale study of savants included a child, Ilene, who knew "practically every song written—who wrote it, what show it is from (or film), who first recorded it, in what year it was popular, etc." Another child in the study could recite the actor who played each part in the TV program *Roots* after once watching a quick display of the credits. John, the first savant Treffert ever encountered, memorized the Milwaukee bus system; if you told him the time of day and a bus number, he could tell you the precise location of that bus.

But those turned out to be almost run-of-the-mill memory exploits. As Treffert recounts in his book *Islands of Genius,* during his decades of working with savants, he encountered savants with skills—and memories—so notable they had the press running three-ring circuses around them. Daniel Tammet, a man with Asperger's disorder, memorized pi to the 22,514th decimal; he recited the figure without error in just over five hours. A pair of identical twin autistic savants memorized every question and answer (as well as what the host wore) from every episode of their favorite game show. An exceptional memory is, as Treffert once characterized it, "integral" to savant syndrome.

Joanne's pilot study had suggested that child prodigies' family members had a heightened attention to detail, a trait associated with autism. She had worked closely with only two prodigies, barely scratching the surface in her quest to understand the underpinnings of their abilities. But already she had discovered that both children had extraordinary working memories—an occurrence highly unlikely to occur by chance. It seemed probable that extraordinary memory was an important characteristic of prodigy—and another possible link to autism, or at least autistic savants.

Chapter 4

Growing a Prodigy

Can you create a prodigy?

If you focus on prodigy's external markers—the astounding work with the brush, the early entrance to college, the excellence at the piano—it *almost* seems possible. Maybe with the right expertise, maybe with enough determination, you could get the right teachers, instill an unstoppable work ethic, and place a kid on the fast track to Carnegie Hall.

Or could you?

Between the summer of 2010 and the summer of 2011, Joanne zigzagged across the East Coast and the Midwest in pursuit of prodigies. Her sample swelled from two to nine. That may not sound like a big number, but it was the largest group of prodigies anyone had assembled in eighty years.

As Joanne went from one home to the next, she examined the kids and spent time with their families. She listened as the parents described their experiences raising their children. Did they share some little-known secret to unleashing prodigious abilities? What portion of the prodigies' skills was the product of nurture, careful shaping in the hands of adept parents, and what portion was the product of nature?

It's a question perhaps best investigated by exploring the lives of two of those nine prodigies: Jonathan Russell, the son of an expert, and Lauren Voiers, the daughter of an amateur.

⌐

Jonathan Russell is a twenty-year-old New York University student with curly dark hair, thick eyebrows, scruffy facial hair, and a near-encyclopedic knowledge of film scores. He recently released an independent album of fifteen original instrumental tracks, gives occasional violin performances in Central Park, and often rescores portions of popular movies, TV shows, and video games for fun.

The pieces that make it into Jonathan's schoolwork or onto his album or YouTube channel are only a fraction of his original compositions. As Jonathan describes it, music is always on his mind: "My brain is in constant music mode a lot of the time. I kind of have this internal iPod, but instead of playing music that already exists, it composes completely by itself, and there's nothing I can do to stop it."

The idea of having an internal music composition system operating on autopilot sounds incredible, but it's something of a mixed blessing. The stream of music gets stronger the closer Jonathan gets to sleep. Sometimes he has to play the music or write it down, just to get it out of his head.

He has a similar, almost reflexive ability to imagine people's voices. When he sleeps, every person he dreams about has a distinct voice. When he reads, he can easily conjure up the sound of a favorite TV character (often those with a British accent, like Stewie Griffin, the evil genius toddler on *Family Guy*) or a friend speaking the words, as if he were listening to a one-of-a-kind audiobook.

Music has been a part of Jonathan's life for as long as he can remember. He was born into a musical household in Riverdale, an upscale part of the Bronx. His mother, Eve Weiss, is a guitarist; the *New York Times* chronicled her 1983 debut New York performance. She stopped performing when Jonathan was three or four, but she's still a full-time guitar instructor; his father, Jim Russell, works in IT but minored in music in college.

Almost from birth, Jonathan seemed interested in sound. He loved to strum Eve's guitar, and when Eve sang to him in the stroller, he dictated song choice with a shake of his head.

When Jonathan was eighteen months old, he pointed at a picture of a violin on a bag slung over a doorknob and said "violin"; he repeated the word whenever he saw an image of a violin in the house or on TV. Soon after, Jonathan picked out the sound of a violin on a recording and again piped up with his new favorite word: "violin." At eighteen months, Jonathan recognized the instrument by sight and sound.

"My reaction was two things. It was, I have to become a Suzuki teacher—that's the best method I've heard for teaching young kids— and let's get him started as soon as we can," Eve said.

Eve began teacher training for Suzuki around the time Jonathan turned two. It's a "mother-tongue approach" to music instruction in which children learn an instrument just as they learn a language—by ear and with parental support. Jim and Jonathan tagged along when Eve had class, and father and son passed the time by strolling the music institute playing a game they called "looking for Mr. Bach." Once, Eve and Jim put a "teeny tiny violin" under Jonathan's chin, just to see what would happen. He got a huge grin on his face.

Eve and Jim played the introductory Suzuki violin CD for Jonathan at home and in the car (Eve tells her students' parents to "play it till you want to throw it out the window"). They put Jonathan to bed listening to Bach's Partita in D Minor on violin. At two, Jonathan's grandfather gave him an old keyboard, and Jonathan tinkered with it for hours at a time, figuring out sounds.

When Jonathan was two and a half, Eve contacted the head of the School for Strings, a Suzuki-based music school in Manhattan. "I said, 'You know, my kid, he can hear the sound of a violin, he loves music, he's very musical, he's left-handed, and he's not potty trained yet,'" Eve said. The director told her that the school didn't take on kids who weren't potty trained. He told her to wait a year. "I kind of said, 'I know my kid, and he's gonna be playing "Twinkle" on the violin before he's potty trained.'"

Jonathan's pediatrician put Eve in touch with Monica Gerard, a newly trained Suzuki instructor, and Jonathan began taking weekly lessons with Monica a few months before his third birthday. In place of a real violin, Jonathan and Eve wrapped a couscous box in paper, painted it, and stuck a ruler in it (a typical technique used when teaching young children to hold a violin); he named it Peter after the title character in *Peter and the Wolf.* Eve already knew how to play a bit of violin ("badly," as she put it), so, as required in Suzuki training, she helped him at home.

Jonathan gave his first performance at three (using a real violin) as part of a recital for Eve's students. His face lit up as he stood in front of the audience, his baggy checked pants held up by suspenders. "Jonathan got up onstage and took his bow and held his violin out, put it under his chin, picked up the bow, played one round of 'Peanut Butter Crackers,' and then he wouldn't get off the stage," Eve said. ("Peanut Butter Crackers" is the name Eve uses for a common introductory Suzuki rhythm.) "He just grinned—stared at the audience and grinned—and they applauded and we had to drag him off the stage."

Jonathan picked up pieces quickly, but his motor skills were underdeveloped. He couldn't hold the violin correctly; he had trouble positioning the bow. It was a struggle for him to bend his arm at the elbow. Sometimes he fell over while practicing posture.

Eve and Jim had been keeping a close eye on Jonathan's development. He was hypersensitive to loud sounds: the rock music at a dolphin aquarium show made him hysterical. Smells, too, like the fumes from a car or the odors of a restaurant, could set him off. He was clumsy, zigzagged when he walked, and had trouble staying in a straight line in tap class. At his violin teacher Monica's suggestion, Eve and Jim had Jonathan formally assessed. He was diagnosed with a sensory processing disorder—a condition in which people under- or overreact to sights, smells, or sounds (or some combination thereof).

His clumsiness made the technical aspects of the violin a struggle, but there were flashes of talent of a different sort. When Jonathan was three or four, Monica asked him to figure out the melody to "Old MacDonald." When he played it for her, Jonathan put in a slide—a little embellishment of sorts. At first, Monica thought he had made a mistake. But when he played it back to her the same way, slide and all, three more times, Monica realized he was doing it on purpose; her pint-sized student was *improvising*.

It was no fluke. About a year later, Jonathan went with Eve's mother to see a klezmer band, a group that played traditional eastern European Jewish music. One of the tunes was familiar to him—Eve's mother had sung the popular "Bei Mir Bist Du Schoen" to him before, replacing the words with "the bear missed the train"—and when Jonathan got home, he got his violin and pieced the song together.

It was impressive, but not a complete shock. "Okay, he's a Suzuki kid, they're trained to do things by ear," Eve remembered thinking. But once he had the song nailed down, he began improvising on it; he left the melody intact but altered the rhythm at certain points, a technique often used by jazz musicians. "That was like, yeah, okay, five-year-olds don't normally do that kind of thing," Eve said. "My mouth kind of dropped open."

A kid with a knack for jazz improvisation was a hot commodity. The next year, Eve took Jonathan, then six years old, to a recital for older children. After Jonathan gave an impromptu performance, a musician in the audience asked Jonathan to sit in with his jazz band, Dulit's Dixieland Devils, at a Tarrytown restaurant. The group's youngest member was a roughly fifty-year-old drummer. Some of the band members were supportive; some complained about Jonathan's technical imperfections. It became a regular gig for Jonathan.

A string of referrals created a domino effect of performance opportunities: at seven, he began playing at Arthur's Tavern, a live jazz joint in Manhattan's West Village ("the oldest gig in New York," as Eve described it), and the Cajun, a Chelsea venue. "I had reservations about the cigarette smoke, but not playing at a bar," Eve said of those early gigs. "It wasn't a honky-tonk kind of drunken bar. It was a jazz club."

That same year, Jonathan began making the jazz festival rounds, where his head barely reached the elbows of the adult musicians. His performances snagged enough attention that the *New York Times* profiled him, commending his "sophisticated improvisations on the melodies of jazz standards." Over the next few years, more friend-of-a-fan referrals put Jonathan onstage with a ninety-one-year-old Les Paul at Iridium in New York City and an eighty-one-year-old Bucky Pizzarelli, a pairing of "the rotary phone and the cell phone" of jazz, at the North Carolina Jazz Festival.

For all his love of performing, Jonathan hated to practice, a characteristic that set him apart from most of the other prodigies. He knew all the pieces, but his motor skills were still behind the curve; he hated when Eve corrected his technical mistakes.

He rehearsed, but only because Eve told him it was one way or the other: if he didn't practice, she would cancel his gigs. He practiced, about two and a half hours a day; neither Jonathan nor Eve could tolerate more than that.

That changed when, around the time Jonathan turned twelve, a friend played him a piece of the score from the first *Pirates of the Caribbean* movie. "I was like, 'Oh my God, I forgot how good this was,'" Jonathan recalled. "I started listening to all the albums, and I kind of memorized everything on the first listen through, which happens sometimes."

He spent a year writing new arrangements for the music. He went to fiddle camp, but instead of playing the usual fiddle tunes, he corralled the other campers into performing *Pirates* music with him. He did the same thing with the *Lord of the Rings* trilogy. He listened to the music, he memorized the scores, he wrote new arrangements for it—for hours at a time.

Instead of having to urge him to practice, Eve finally told him to stop; she couldn't take listening to *Pirates* music anymore. She urged him to compose his own stuff. Jonathan researched top-of-the-line composition equipment on discussion forums and Web sites and, using the money he earned from his jazz gigs, bought a new computer for composing.

Sometimes he sat at the piano and improvised, but for the most part songs came to him fully formed. During high school, he got up at 5:30 a.m. to compose because it was a time when he felt particularly inspired.

Jonathan set his sights on film scoring. Eve asked around about composition lessons, and a friend put her in touch with an NYU composition professor who took a thirteen-year-old Jonathan on as a student. Jonathan developed a huge knowledge bank of scores and score trivia. During one conversation, he noted that Hans Zimmer wrote parts of the score for the first *Pirates of the Caribbean* movie but his contribution is uncredited; Peter Jackson knew the composer Howard Shore had found the theme for Rohan in *The Lord of the Rings: The Two Towers* when he started humming it in the car; in *Batman Begins,* Zimmer composed an incomplete theme for an incomplete human being. Jonathan began an annual tradition of creating a medley of the best original score Oscar nominees and posting it online.

The performances continued, often influenced by Jonathan's fascination with film scores. There were jazz festivals throughout the United States (he "pluckishly improvised—using the Lone Ranger's theme" at the Sacramento Jazz Jubilee) and a music tour in Hungary (he watched *Star Wars* in Hungarian). Jonathan put out a few self-produced CDs, all of which included his improvisations. He played with Wynton Marsalis at Rose Hall at Jazz at Lincoln Center as part of the Nursery Song Swing concert series when he was thirteen. At fifteen, he performed the improvised violin segment of *One Night with Fanny Brice,* an off-Broadway production, three nights a week—"sprightly contributions," as described by the *New York Times.* But Jonathan's heart was in composing movie scores—a path that Eve, herself a classical musician, had never envisioned for him.

"We did have to make him practice; that's what he hated doing about everything. But the improv came from him. Nobody could have taught this kid to improvise when he was younger, and the composition end, which he owns even more, that was the driving force behind him," Eve said. "We couldn't have made him compose, we couldn't have made him sit for hours like he did, but we made him practice."

⌐

Lauren Voiers grew up in Westlake, a suburb on the west side of Cleveland. She's five feet eleven inches with long hair—sometimes blond, sometimes brunette—a round face, and caramel-colored eyes. Her father, Doug, is a cosmetic dentist, her mother, Nancy, a nurse turned stay-at-home mom who has now returned to nursing.

From the time Lauren was two or three, she had, as her father put it, "a very, very, very, extremely strong desire" to create. "Sometimes I would come home from work, and my wife would brace me for the carnage that had occurred at the home," Doug said. "She did destroy good parts of our home over the years."

At three years old, Lauren got her hands on a permanent marker and drew on all four walls of a bedroom, "broad strokes, as high as she could reach," Doug recalled; the Voierses had to tear the wallpaper down. She drew on the carpet until it had to be ripped off the floor. She carved designs into the woodwork, once etching into a custom-made window seat in her bedroom.

As she got a bit older—four, five, six—Lauren's creative urge persisted, but she channeled it onto more traditional surfaces. She drew faces and objects in great detail without looking at any sort of reference material. She painted by number; she painted on backpacks and clothes. She assembled jewelry from kits, made a jewelry box out of clay, and poured colored sand into bottles. She had a knack for making posters that won her a couple of school contests. In middle school,

Lauren grew interested in architecture and sketched out designs on graph paper, creating modern, angular homes and geometric spaces.

Nancy and Doug both enjoyed the arts—Nancy played the piano; Doug took a ceramics course in college and dabbled with watercolors—and they bought Lauren markers and crayons and other art supplies on Christmas and her birthday to support her interest. Art, they thought, made for a great hobby.

In seventh grade, Lauren's eye wandered to her dad's painting supplies. Inspired by one of her dad's art books, she borrowed a canvas and some paints and tried to replicate a Thomas Kinkade painting of a river running through a forest. Over the next few months, she produced a few other small landscapes—a couple of houses, another nature scene. She painted a couple of times a week for four or five hours at a time. "Eventually, it got to the point where she was producing things that were fairly amusing," Doug said. "But of course, we wound up with four kids, so we're making babies, and my wife's a critical care nurse, and I'm a cosmetic dentist and building a business, and we're living our lives, so we're not paying much attention to the quality of what she's doing."

When Lauren was thirteen, she saw Marla Olmstead, a then-four-year-old artist with big eyes and chin-length hair, on *The Jane Pauley Show*. She was riveted by the girl's story and even more so by her paintings. "It was filled with a lot of micro-detail and smaller areas," Lauren said of one of Marla's works, a large, fiery piece punctuated by dark splotches. "It was kind of 3-D, kind of like creepy in a way."

Lauren sought out more images of similar artwork. During breaks at school, she went to the library and studied the abstract paintings on artists' Web sites. When she got home, she did the same thing, poring over artists' Web sites and examining art books until she was spending two and a half hours a day inhaling art. "I absolutely became obsessed," Lauren said.

She abandoned the landscapes she had been painting and tried her hand at abstracts ("using my fingers, using my hands, just kind of experimenting with paint"). She quickly moved toward cubism, depicting objects as composed of—and alongside—an array of geometric shapes or, as Lauren put it, "breaking things down to their simplest form."

She hustled to the art room during lunch and free periods; she stayed after school to paint. The school art teacher called Doug and Nancy after she saw Lauren's early abstracts. She thought their daughter had a gift. After that, the Voierses bought Lauren paints and the large canvases she wanted. She completed fifteen, some as large as three by four feet, that year.

Her production accelerated once she started high school. She spent six hours a day painting, then seven, eight, nine hours a day. Her other activities—tennis, basketball, schoolwork—fell by the wayside. She pieced together a couple of hours of painting at school; study hall, lunch, any extra time she had went toward her artwork. But she did the bulk of her painting at night, after everyone had gone to bed. Doug converted their attic space into an art studio for Lauren during her junior year of high school, and Lauren stayed up past 3:00 a.m., past 4:00 a.m., sometimes not sleeping at all before school and then crashing during first and second periods. "I went kind of crazy on it," Lauren remembered.

Her artwork again spilled over onto the walls, this time the walls of her own room, which she decorated according to a different theme every year—sophisticated jungle, "hippy dippy trippy" murals, metallic green with graffiti.

There were a series of victories. Her painting *Sisters,* a crimson-and-orange work in which two girls appear to be embracing ("showing the love I have for my sister"), was a regional finalist in the 2006 Ohio Governor's Youth Art Exhibition. The next year, *Transparency,* an autumnal-colored piece in which a woman is visible among an array of shapes ("a more stained-glass effect . . . with many layers and dimensions"), won Scholastic's National Gold Key Award. A friend's mother commissioned her to paint a landscape of their house. A Cleveland art dealer sold a few of her pieces, including *The Cellist,* a nine-by-four-foot bronze-and-violet cubist piece that spanned three canvases ("one of the best paintings I ever made"). To the Voierses, though, art still felt more like a hobby than a viable career. "Everyone knows artists are starving," Lauren said. "The odds of becoming an artist that can actually make a living at it is—I don't know what the odds are, but they're ridiculous."

A phone call changed everything. When Lauren was seventeen, a California art agent contacted her parents. He had seen an image of *Sisters* that Lauren's mother, Nancy, had posted online. He wanted to take Lauren on as a client. "Our first conversation on the phone, my wife got all excited and I said, 'Okay, I want to talk to this guy, because he's full of it,'" Doug said. "I was very, very worried for her going into that world. My wife was excited about it, about the prospect, and I was very dubious and I was skeptical."

The agent proposed a trial online auction, just to see how things went. He sold around $179,000 worth of Lauren's art; Lauren's cut amounted to more than $40,000. Lauren signed a representation agreement with the agency a couple of months later on her eighteenth birthday. Her parents contacted the school to help her set up a reduced schedule for the last semester of her senior year so she could travel to art events.

"Right after that, it was bam, bam, bam, bam, bam," Lauren said. She zipped off to shows and auctions across the country, making her way to New Jersey and New York, La Jolla and Las Vegas, Virginia and Maryland. "It's something I had been dreaming about," Lauren said. "Every time I would be working on my art, I'd have it in the back of my mind, imagining, what if I'm painting this for a show? I always dreamed it would be going somewhere else besides sitting in the attic." Her price tags shot upward, with her large canvases eventually selling for around $20,000 each.

In the midst of her immersion in the professional art world, Lauren graduated from high school. She rented studio space in Cleveland—a loft downtown with big windows, hardwood floors, and redbrick walls—and then in a church on the east side where she worked in a converted space in the rafters. When she met Joanne in September 2010, she was officially a professional.

"Other people began recognizing her talent far before we did, mainly because they had the context," Doug recalled. "So here they are comparing what she is doing to all of these other children of the same age-group. We didn't really have that ability to compare; we just saw what she was doing."

―

Jonathan's and Lauren's families aren't polar opposites. Both kids come from financially stable, two-parent homes. Both families provided their kids with the supplies they needed to pursue their interest— a violin for Jonathan; paints and canvases for Lauren.

But their families weren't exactly following the same playbook, either. Lauren was almost entirely self-taught; Jonathan took violin lessons, piano lessons, and composition courses. Lauren's family knew little of the art world; as she put it, "The whole art business— that was brand-new to all of us." Jonathan's parents knew the ropes. "I'm a musician," Eve said. "A lot of the parents of prodigies aren't musicians or aren't this or aren't that and are kind of lost, but my thing was, I'm a musician, I know what all the pitfalls are. I was gonna make it much easier for him to succeed than it was for me."

The question of how significantly parents contribute to their children's achievements is an old one, and it's one that prodigy parents often face. While many families seem bewildered by their child's advanced abilities, there have, historically, been parents eager to take credit for their children's achievements. In the 1910s, for example, a small group of parents proclaimed that they had turned their children into prodigies. One of these parents, Leo Wiener, a Harvard professor whose children included Norbert Wiener, a famous prodigy and eventual MIT mathematician, claimed that his children were "not precocious," "not geniuses," and "not even exceptionally bright." "I could take almost any child and develop him in the same way," Wiener said. "It is merely the method of imparting learning."

In other words, these parents claimed that their prodigies were normal children. They had no inherent advantage in the smarts department. In the parents' telling, these children excelled because their education began at a young age—perhaps at two or three—well before the age at which formal schooling typically began. Other common threads among the parents included rejecting baby talk; introducing letters, numbers, languages, and other areas of study at a young age; and making learning interesting. According to these parents, their children's outcomes were the product of their parents' efforts.

The view that parents could create a prodigy out of almost any child is an extreme nurture take on prodigy, one that bestows on parents great control over their children's minds and abilities. This nurture-oriented perspective is oddly reminiscent of some of the darkest moments in the history of autism. For decades, scientists and the public debated whether *parents* might be to blame for their children's autism.

The groundwork for this debate was laid early. In his earliest writings on autism in the 1940s and 1950s, Leo Kanner entertained the idea that parents might have played some role in their children's condition. In particular, he questioned the impact of two practices he believed were common among the parents of the autists he saw: "stuffing" the children with "verses, zoologic and botanic names, titles and composers of victrola record pieces, and the like" and subjecting them to "emotional refrigeration." He eventually concluded, though, based on the fact that his patients exhibited autism symptoms almost from birth, that autistic tendencies must be innate: the environmental influence, he wrote, was "not sufficient in itself" to cause autism.

Others disagreed. Bruno Bettelheim, the vocal, long-term director of a children's residential treatment center at the University of Chicago, was convinced that autism was a product of nurture. He believed that autistic children internalized the perceived negative emotions of their closest caregivers—most often, their mothers. Drawing on the tale of Hansel and Gretel, Bettelheim claimed in his 1967 book, *The Empty Fortress*, that an autistic child would see his or her mother as the "devouring witch" of the fairy tale and that his or her withdrawal was a defense mechanism against the mother's perceived "destructive intents."

Eventually, scientific research, particularly twin studies, swept the nurture conception of autism aside. In the first of these, a 1977 study, the psychiatrists Susan Folstein and Michael Rutter found that among sets of twins in which at least one of each pair was autistic, the second twin was far more likely to have autism if the twins were identical than if they were fraternal. Their finding suggested a strong genetic component to autism because the identical twins had more DNA in common with each other than the fraternal twins (nature), but all the twins shared a prenatal environment and a home environment (nurture). These findings were buttressed by other twin studies and by studies that identified a higher prevalence of autism-related traits in autists' family members—a finding that suggested a genetic basis for such traits. It has since become conventional wisdom that autism has a large genetic component. In most parts of the world, the idea that parents are to blame for their children's autism has been tarnished and cast aside.

But what about child prodigies? Are their behaviors, too, largely the product of genetics? After spending more than a decade investigating prodigies, David Feldman thought there was an innate core to prodigious skill. He and Lynn Goldsmith concluded in *Nature's Gambit* that all six of their subjects had "striking and extreme" talents—abilities with which they were born. "If these children themselves were not truly gifted," they wrote, "they would not emerge as prodigies."

There are occasionally cases in which a child's story can serve as an almost undeniable example of such extreme innate talent. Kelvin Doe, for example, grew up in Freetown, Sierra Leone, with few resources; he's the youngest of five children and was raised by a single mother in a community that had electricity only once a week. At eleven, he began rooting through trash heaps for scrap electronic parts. He used what he found to create a battery to power the lights in his house. The self-taught inventor later assembled a generator and an FM radio transmitter. He's since been invited to visit MIT, given a TEDxTeen talk, and signed a $100,000 contract to develop solar panel technology in Sierra Leone.

It's easy to imagine, though, that other prodigious children never fully develop their talents. According to Feldman and Goldsmith, while innate talent was the engine for the achievements of the children they studied, most prodigies could never reach their potential without catching a few breaks. They explained that even children endowed with magnificent abilities benefit from significant familial support and superb teachers and from choosing a field both valued by society and conducive to the rapid development of expert ability. Only when all of these factors work in unison, in "a beautifully choreographed co-incidence of forces," can a child's full potential be revealed.

It's an insightful conclusion; it illuminates the circumstances under which the abilities of a prodigiously talented child might fully develop. It suggests that even if prodigies have a baseline innate capability that can't be taught, most still need a certain environment to maximize their talent. But providing that environment, figuring out how to raise children who, as one reporter put it, "stand out like Gulliver among the Lilliputians," is a challenging parental task.

These families find themselves unexpectedly confronted with a child's ravenous need for new information. They scramble to find materials or teachers to satisfy and engage a mind with a seemingly endless capacity to learn. It's an urgency-infused struggle; the kids often seem to *need* to develop their skills in the same way that they need to breathe (the music prodigy Jay Greenberg once told his mother that if he wasn't composing, he would be dead).

So how *do* you raise a prodigy? It's not an easy question to answer. Among the prodigies' families, there's relatively little consensus on some of the toughest questions.

How, for example, should you educate a prodigy? Lauren and Jonathan followed similar, and relatively traditional, education tracks—Lauren graduated from a public school, Jonathan from Manhattan's Professional Children's School—and both followed a normal grade-level progression. Two of the music prodigies, on the other hand, rely on homeschooling, an option that gives them schedule flexibility, lets them set the pace, and allows for a deep dive into interesting subjects.

William did a few years of mainstream education at grade level while pursuing his "real learning" at home, but this turned out to be a short-term solution. At six, he got fed up with what was, for him, a simple and tedious multiplication test, and he refused to complete it. Instead, he handed in a test he made himself in which he calculated the square roots of the numbers 1 through 10. For the more difficult numbers, he wrote out the answer to the sixth decimal place. "He then put his pencil down and said 'I'm NOT doing this anymore!'" Lucie recalled. The school promoted him to sixth-grade math, and his teachers gave him seventh- and eighth-grade math exercises to keep him busy.

There's a media question, too. The press loves a child prodigy, but how to handle the interview requests? Partly due to the way in which Joanne selected the first batch of prodigies—searching news stories for reports of children with preternatural abilities—many of the kids were media savvy. They gave interviews freely and had their own Web sites; they eventually had Facebook pages and Twitter accounts.

Some of the coverage resulting from prodigy interviews is warm and fuzzy and a way to further the career of a child whose inner engine has been shifted into high gear since birth. But there are risks to media exposure, too. There's always the chance that an interview will lead to a less-than-favorable piece. Even if the initial coverage is kind, once the information is out there, it can be rehashed, recut, and reconsidered by people to whom the family never entrusted their story. For the more prominently featured children, there is also the prospect of being forever followed, forever documented—even if they no longer want any part of the limelight.

It was only later in Joanne's career, once her research began attracting journalists' attention, that she heard from media-shy prodigy families, people who wanted to help advance her research but didn't want a reporter anywhere near their home. Alex and William's parents, for example, shielded their boys and their family from the press. "I think we'll just avoid all that and leave it as so," Lucie said of interview opportunities. "We love our quiet lives that are undisrupted."

These parents often face unexpected financial burdens (who budgets for professional art courses for a teenager?) and social struggles (what does a kid who studies theoretical physics talk to other eight-year-olds about?). There are as many answers to these questions as there are prodigy families. On the social front, for example, some seek out potential common ground the children may share with their age-mates; others focus on nurturing relationships with siblings; still others try to help ease their children into environments that offer common ground with intellectual (but far older) peers. Often the same family will try out different approaches as time and circumstances change. There is, for better and for worse, no single way to raise a child, even if that child is a prodigy.

⟶

While parental capabilities and resources and the home environment in which the kids are raised may impact the prodigies' ultimate level of achievement, these factors don't explain the *source* of what Feldman and Goldsmith described as the prodigies' "striking and extreme" talent.

Despite their parents' differing levels of expertise in their fields, Jonathan and Lauren share a couple of traits that seem essential to their abilities. Both walloped the working memory section of the Stanford-Binet IQ test, just like the other prodigies. And once they discovered their areas of specialty, both Jonathan and Lauren demonstrated a dizzying need to pursue that interest.

It's this passion that seems to drive the prodigies. It's a popularized principle of psychology that it takes ten thousand hours to become a world-class expert in something, and the prodigies' insatiable passion explains how they might manage to get their ten thousand hours in by such a young age. In Lauren, the source of that passion was clearly internal. For most of her teenage years, her parents didn't think an art career was realistic; they were happy to provide Lauren with supplies, but they wanted her to focus on school. Jonathan's case looks somewhat different. His mother knew the ins and outs of a music career. Eve encouraged Jonathan, though for a time she could only cajole him into practicing a couple of hours a day. But when Jonathan discovered film scoring, the area in which he is truly prodigious, he demonstrated an *internal* "rage to master."

Every parent Joanne met similarly insisted that the determination to paint, play, compose, or study originated with the child. Most of the parents recount their children seeking out endless information, music practice, or time at the easel *despite* their parents' pleas that they go outside, run, climb, dance, or play soccer. Terre Grossman worries about Greg's falling asleep with his cell phone in hand and waking up to answer business e-mails. Lucie gets William out of the house by baiting him with chalk; if he is going to write equations, at least he'll do it outside.

So where do the prodigies get their relentless drive? The answer may be a connection with autism. Just like extraordinary memory, this tendency toward obsessive, almost all-consuming interests is another trait the prodigies share with autists and autistic savants.

Chapter 5

The Evidence Mounts

When Richard Wawro was young, it would have been absurd to think that he might become an artist.

He was born in Newport-on-Tay, Scotland, in 1952, with cataracts in both eyes. He had them surgically removed when he was only a few months old; even afterward, Richard could see very little.

His vision problems were in some ways the easiest to handle. Richard rarely slept; his father, Ted, a former Polish army officer, would later recall that Richard never slept more than two hours at a stretch as a young child. He was prone to violent tantrums: he threw himself on the floor, where he floundered, kicked, and let off high-pitched screams. He walked in tight circles and spun objects for long periods of time. He could tap a single piano key for hours. "It was enough to drive you mad," his father later told a reporter. Ted eventually dismantled the piano; he used the wood to build two coffee tables.

When Richard was three, his parents took him to a nearby hospital for testing and were told that their son was moderately to severely retarded. The doctor advised them to institutionalize Richard and forget about him, just as he had done with his own son. "My mother was a fierce mother, a very protective mother. She would never put Richard in a home," Richard's younger brother, Mike, a technology consultant in Edinburgh, recalled. Richard's parents kept him at home. Ted worked as a commercial librarian, and Olive, his mother, returned to teaching.

Richard spent his days at the home of a neighbor, Mrs. Jean Currie, where he discovered her chalk and blackboard. Mrs. Currie never expected Richard to use it, but one day Richard drew a fireplace that resembled the one in the Curries' home. When Mrs. Currie told Ted what Richard had done—that Richard had *seen* the fireplace in her home and drawn it—Ted dismissed it as wishful thinking. But Richard continued filling the chalkboard with images; his ability to distill what he saw into drawings became undeniable.

As Richard got older, Olive tried to enroll him in school, but she struggled to find one that would accept him. Eventually, she contacted Molly Leishman, a special education teacher who agreed to work with the golden-haired six-year-old on a trial basis.

It wasn't easy. Richard was restless. He screamed loudly; he spun wildly. Richard loved music—he often drummed on desks and chairs—but he grew hysterical if a teacher played a record he disliked. As Molly later recalled, Richard "was never still, unable to see properly or to speak, and always I felt he suffered."

At some point, Molly noticed that flickering light entranced Richard. It had done so since infancy. As soon as he could lift his head, he had turned to look at the sun. As a toddler, he had stared incessantly at a pool of water reflecting light. Even in photographs, his face was often tilted toward the light. Drawing on this interest, Molly covered mobiles, building bricks, and other toys in iridescent paper. Richard happily sat at his desk and watched the shiny paper glisten.

As Richard began to settle down at school, Molly tried something new. She pinned paper to his desk, kneeled down next to him, and scribbled a few lines with a large red wax crayon. Richard was interested. He refused to hold the crayon himself, but Molly placed the crayon in his hand and guided him through making a few marks. He quickly began scribbling on his own, keeping his eyes close to the paper.

It became his favorite activity. He scribbled away, happy and content, producing pictures on par with those of his classmates. It was great therapy for him, Molly thought. She hung his pictures alongside those of the other students.

Then, without warning, Richard's drawings changed: Molly found Richard not just scribbling but creating an impressionist picture. "It wasn't the usual picture, you know, with a bowl for the head and a bowl for the body and the four little matchsticks for legs that you sometimes get, and you think it's wonderful if you get from a handicapped child. It wasn't anything like that *at all*," Molly recalled in a documentary about Richard. "He was doing something that even a normal child of his age, a child that wasn't handicapped at all, couldn't have done."

Richard and his crayons became inseparable. He used them to draw almost everything he saw: his breakfast, the school bus, boats, trains, buses, and the cartoons on TV. He devoured the paper his family provided to him. He filled sketchbooks full of blank pages in a day or two. When he ran out of paper, he drew on the wallpaper, punctuating it with images of Yogi Bear and Huckleberry Hound. "He was exploding art out of his fingertips *every* day," Mike said.

These early drawings already radiate a sense of motion: Richard captured boats mid-sail; the roiling of a turbulent sea; a kangaroo hurtling forward. Unlike the flat, two-dimensional images usually generated by young children, his drawings have a sense of depth and perspective. The cartoons Richard drew for Mike—lively, vibrant, colorful depictions of the black-and-white shows that were on TV— jump off the page.

Richard still loved repetition. As an adolescent, he sometimes woke in the middle of the night and turned on the radio to listen to the test signals—pulse-like beeps that repeated every minute. If Mike turned it off, Richard would switch it back on. He still craved consistency; he became inconsolable when his favorite cereal bowl was lost during a car trip. He remained transfixed by light and could gaze at the colors cast off by a spinning prism for long stretches.

But the more Richard drew, the more he relaxed. Over the next few years, he steadily built his library of pictures to include depictions of animals (*A Crocodile; Bees Swarming; Cats at Play*), fairy tales (*The Witch from Snow White; Robin Hood 2*), and transportation (*Tractor; Train Through the Hills; A Plane at Sunset*), and those were only the tip of the iceberg. The screaming fits, the sleepless nights, and the extreme withdrawal dissipated.

Richard had begun speaking late; even then, he could form only a few words, he struggled to articulate clearly, and his words came out as a whisper. But when Richard was eleven or twelve, he had a breakthrough. He transitioned, suddenly and inexplicably, from whispering to speaking at a more typical volume. His pronunciation and vocabulary, along with his ability to string words into full sentences, were still fairly poor, but from that point on at least other people could hear him.

When Richard was twelve, Marian Bohusz-Szyszko, a Polish artist and art instructor, visited the Wawros to examine Richard's artwork. Bohusz-Szyszko looked through Richard's collection, which already included hundreds of drawings and paintings, and watched as Richard banged out eight new drawings in forty-five minutes.

"He was gesticulating and gasping. He was really theatrical," Mike said of the professor. "He was almost breathless at some of the stuff." Bohusz-Szyszko was particularly intrigued by Richard's bold, broad stroke outlines and the way he applied layers upon layers of crayon to his pictures. He urged Ted and Olive to help Richard develop his talent. When he later wrote up Richard and his artwork in a Polish-language newspaper, he said that Richard's work left him "thunderstruck." His drawings of machinery—cranes, ships, motorcars—were rendered "with the precision of a mechanic and the vision of a poet."

As word of the nearly blind artist spread, Richard's art, which had been a family matter, became a public phenomenon. People began showing up at the Wawro home to test Richard and to try to figure out how he was creating his spectacular drawings. The experts tagged Richard with an IQ of 20 or 30.

"I thought, that's complete BS. That's just not Richard. I was twelve at the time, and I could tell you that you're wrong," Mike recalled. "Well, what it means is your tests aren't right for him. It doesn't mean that Richard's stupid. The tests gave the wrong results. They were of no relevance or interest to him."

Mike and his parents saw what the tests could not. They knew that despite the difficulty Richard had communicating, he showed occasional flashes of insight—he grew upset over the Vietnam War; he was very thoughtful about death—that hinted at his depth of understanding. There was a lot going on in his head, his family felt; he just couldn't articulate it.

A few years after the visit from the Polish art professor, Richard was "discovered" for the second time. Richard Demarco, "the Edinburgh impresario," a Scottish artist and art promoter, visited the Wawro house and examined Richard's drawings. He, too, was amazed at what he saw, and he set up an exhibition of Richard's drawings at his Edinburgh gallery. "This is them really being properly framed for the first time, properly lit in a gallery, people going around with glasses of wine saying, 'Oh this is fabulous,' and all the rest of it. From that point on, there was self-recognition that he was an artist," Mike said.

It was eye-opening for Richard's parents, too. Ted and Olive had given away a handful of Richard's drawings before, the odd cartoon or animal drawing, to friends or neighbors; they had sold a few of his pictures to raise money at church fund-raisers and bake-offs, but this was the first time that Richard's works were sold in a professional setting. "On the first evening Michael ran up to me and said, 'Dad, a picture was sold!'" Ted later told a reporter. "Someone had paid £16 for a drawing done by my mentally retarded son. I felt like crying."

Afterward, Ted organized a flurry of exhibitions. Before the end of the year, he arranged events at a YMCA in London, Gallery K in Cupar, Scotland, and the Polish Invalids' Club in Edinburgh. At some events, they displayed around three hundred of Richard's drawings; at the exhibit at the YMCA, Richard's pictures were stacked three rows high with a mere inch or so of space between them. "My dad was not fussy at all; any exhibition was good," Mike said. "His strategy was to show everything that Richard drew, and he would show it anywhere, at any time." By his family's count, Richard drew 131 pictures that year. The next year, he produced 220 pictures; the year after that, he drew 224.

The vast majority of Richard's pictures were in color, but he experimented with black and white, especially when depicting a historical subject, as he did in *Winston Churchill, 1939,* a strikingly detailed portrayal of the British leader peering over a wall, his country similarly poised on the brink of war. He stuck to the same limited color palette for *Austin in the 1930s,* a head-on street scene in which Richard has blurred the edges of the picture, evoking a faded photograph or soft focus.

Many of his pictures were landscapes, often luminous depictions of light on water. These landscapes were reflections of images Richard had seen in real life or in books, but he rearranged the composition, adding or removing things as he saw fit. He also occasionally flirted with the wacky and the absurd: he drew a steaming mug floating on the cloud line in *Japanese Tea Ceremony* and musical stuffed animals in *Teddy Bear Pop Group.*

At times, the workers at the adult day center Richard attended suggested that he was *too* obsessed with drawing. They urged his family to put more emphasis on helping him learn life skills.

Richard did other things—he loved going to the bookshop and staring at the images in the books (he would select books with pictures and, as one reporter observed, "hold them up within an inch of his face, staring at every flower and every detail of the pictures, sometimes for more than an hour"), he enjoyed playing basketball—but he was always eager to return to drawing. It was what he wanted to do.

As Richard's catalog expanded, so did his reputation. Ted, who had a knack for publicity, never hesitated to invite big names to open one of Richard's events. In 1973, when Richard was twenty, Margaret Thatcher, then the secretary of state for education and science, opened his exhibition at the Polish YMCA Gallery in London. Richard grinned as his father showed the future prime minister his drawings.

At some of his exhibitions, Richard demonstrated his technique. He began each picture by peering closely into his tin and finding a black Caran d'Ache crayon. With his head tilted and his face (and thick glasses) inches from the paper, Richard outlined the subject of his drawing in black, often shaping large figures with a single, confident stroke.

He drew with his left hand, coloring his paper with short, rapid-fire bursts of movement—he jolted his hand back and forth in a frenzy of creative energy while clutching several other crayons in his right hand. To fine-tune the shading and perspective, he applied layers of crayon—often half a dozen or so. He never sharpened his crayons; he used the edges to create fine lines. He colored every speck of paper, and he completed every drawing he began. At the end, he buffed the picture with a cloth to give it a sheen.

Richard showed his parents every picture he completed, and over the years they developed a celebratory ritual. After Richard finished a picture of an Australian resort, a glistening depiction of a series of huts on the beach, he clasped one of his hands with his father's, his fluffy brown hair reaching an inch or two above his father's head. The two held their joined hands over their heads and swung their arms back and forth in a series of big, exuberant gestures—Richard sometimes jumped a bit as their arms reached their highest point—while Ted cheered in Polish. At the end, Richard pointed to the recently completed picture. "London to Greenland," he said, substituting "Greenland" for "Australia" in the excitement of the moment. Ted repeated his words back to him, and then father and son embraced. "Clever boy," Ted said, beaming as he patted Richard's back.

Once the ritual was completed, Richard showed no possessiveness toward his pictures. He was eager to move on to the next one; he was happy to give them away.

⟶

In the late 1970s, the family received a letter from Laurence A. Becker, then a graduate student in Maine interested in gifted education. Laurence, a man who bursts with stories and tells them with a faint southern twang, had seen a short documentary about Richard and asked to borrow a copy of it. Soon afterward, he asked if he could exhibit Richard's pictures at a mini-conference he was organizing on creativity for the gifted and talented. "My evil mind thought, wouldn't it be neat to have an art exhibit by someone who missed the club by 100 points, because, to be gifted and talented, you have to have 130 IQ?" Laurence recalled.

Ted saw an opportunity. He packed up fifty or sixty of Richard's pictures and hopped a transatlantic flight. He showed up at the New York conference, and Laurence exhibited Richard's drawings. No one guessed that the artist was (technically speaking) *not* among the gifted and talented.

The next year, Laurence and his wife visited the Wawros in Scotland. Their house was stuffed with Richard's artwork. There were pictures in the bedrooms, the living room, the dining room, and even in the stairway. There were hundreds more stashed in the attic. Ted eventually cut a small door into the wall in Mike's room to access another portion of the attic where they could keep more pictures. "I can't get over the number of 'em," a filmmaker later said while sifting through piles of Richard's pictures.

Within minutes of meeting Richard, Laurence was convinced that his diagnosis was wrong: Richard wasn't mentally retarded. He *was* gifted. He *was* talented. He was *autistic*.

Autism became the official diagnostic stamp on Richard's condition. It was the label used in news articles; it was the label eventually used in his obituary. It accounted for his repetitive behaviors, his insistence on sameness, and his communication difficulties, and it meshed with his passion for drawing. The family even thought they could identify a contributing factor: Olive had contracted rubella during pregnancy, which can impact fetal development and may increase autism risk.

But there was another set of behaviors—a whole other side of Richard—that Mike thought didn't fit with the popular conception of autism. Richard loved to socialize and dance (he could boogie: while listening to music at his brother's flat in Glasgow, he bounced, swung his arms, snapped his fingers, and rocked out on an imaginary drum set). He enjoyed the social aspect of weekly Mass and developed a thirty-year friendship with a clerk at the local bookstore. He was warm and charming and had a great sense of humor; he was a big hugger.

He was also very emotional, and he often channeled his feelings into his art. When Richard's mother, Olive, died, Richard drew a seaside landscape, *Pembrokeshire Coastline, England*. It's an image of a small white lighthouse perched on the edge of a cliff. The bright blue sea is mostly peaceful, though the surf kicks up a bit at the base of the cliff. The lighthouse is dark, but the sun, somewhere, barely out of sight, shines through a hazy sky.

Laurence's visit with the Wawros in Scotland changed his life. Richard's talent, he realized, was amazing; it was almost shocking. The night he got home, he had a dream that he was walking with Richard and his mother down a sidewalk at the University of Texas. In his dream, Richard took off running—something Laurence had never seen him do—then launched into acrobatics. "I screamed, 'Do you see that?' And nobody in the whole place saw it but me, but I knew what I'd seen, and I knew I had to tell what I'd seen," Laurence recalled. After that, Laurence and a team he put together made a documentary about Richard, *With Eyes Wide Open*.

Laurence estimates that he organized ten U.S. exhibition tours for Richard over the course of two decades. He and Richard, accompanied by either Ted or Mike, jaunted across the country to shows in

Louisville, Austin, Orlando, New Haven, Houston, and Atlanta. At these events, Richard could sometimes be found gazing out the window with his binoculars or closely examining photographs or pictures. Other times, he basked in the attention. "He would revel in the moment. People were coming up to Richard and shaking his hand, 'I like it, I like it very much,' taking photographs with him, and all this kind of stuff. He was acting the star, and he liked it. He'd be smiling his head off," Mike said. "Although he wouldn't be able to say a whole lot or communicate particularly well, he was clearly enjoying it."

Signs of his autism persisted. He still sometimes walked in tight circles. He still loved routine. Every year, he drew a picture of a sunrise as an optimistic symbol of what was to come. Annual milestones like Christmas, Easter, and the summer holiday were very important to him; if his routines were altered, Richard grew agitated.

He never learned to read or write much; for most of his career, when someone asked him to autograph a picture, he flipped it over and drew a sketch on the back. He knew how to form letters, but when he drew signs in his pictures, he often used squiggles in place of words.

His parents handled his daily living needs. Richard was diabetic, and his father gave him his insulin shot every morning and tested his urine twice a day for sugar.

But exhibitions built Richard's social skills, and the recognition he received as an artist built his self-esteem; he eventually began calling himself "an international artist." After a demonstration in New Haven, he held up his drawing and declared it a "world champion picture."

At home, Richard attended an adult learning center. Every day, he put on his coat, loaded up his crayons, and waited for the bus. The bus driver, a person Richard got to know well over the years, was a special figure in his life. But as time passed and the adult learning center lost funding, the programs and facilities deteriorated. Eventually, Richard stopped going. He spent his days at home with his father and his step-mother, a Polish woman who spoke little English. Richard's world shrank further as travel grew difficult for Ted and he could no longer escort Richard to distant exhibitions.

The quality of Richard's drawings deteriorated. He applied layer after layer after layer of crayon—more layers than he had ever used before—until the wax began flaking off the paper. Eventually, he stopped drawing. In 1998, the year Richard turned forty-six, the family stopped naming and numbering each of his pictures.

In 2002, Ted died of cancer. Richard's life, for the first time in years, grew tumultuous. Mike, long convinced that Richard was more capable than anyone realized, fought to have Richard live in a primary care facility near his own home. After a few visits, Mike helped Richard move in. For Richard, it was a massive lifestyle change: the staff expected Richard to exercise much more independence than he had in years; he was called on to decide which clothes he wanted to wear and to help make decisions about his future.

Something inside of Richard opened up again. For a brief time, he returned to drawing. He produced beautiful, rich, intricately detailed creations. He drew a haunting night scene in which a lone boat, illuminated from within, drifts on the water. It's a darker piece than most, done almost entirely in shades of blue, but in the corner the water glistens, reflecting the moonlight. When Richard finished the piece, he returned to his celebration ritual, clasping hands, this time with Mike's daughters, and raising them high.

In November 2005, Richard was diagnosed with lung cancer. The doctors said they didn't expect him to last through the night, but Richard, who had already survived another bout with cancer at age five or six and a life-threatening blood infection in the United States, lived another few months. "Despite what appeared to be a very frail soul, he just had this tremendous life force," Mike said.

He died in February 2006. He was fifty-three years old.

The mysteries surrounding Richard Wawro are many. Richard couldn't perform simple calculations—if you asked him to add two and three, the answer eluded him—but if the calculation had to do with the calendar, like how old someone would be in a given year, he could answer almost instantaneously. If you gave him a date, he could tell you what day of the week it fell on.

He had a stunning visual memory. He never used reference material when drawing; he relied solely on memory to craft intricately detailed scenes—some of these images or landscapes he had seen recently, some he had seen months before. If you pointed to any flag, Richard could tell you which country it belonged to.

His auditory memory was no less impressive. He could differentiate between the patterns of sound made by different trains traveling on different tracks and reproduce the various rhythms by beating them out on a table. He loved music, and when Mike bought him a CD with 1960s pop songs on it, Richard could name the tune, the singer, and the year the song was released after hearing only a few notes. He could do the same with countless other recordings.

No mystery was as great as that of his artwork. He had limited vision and drew with the upper half of his body hunched over a table, his eyes inches from the paper. He'd never taken a formal art lesson, yet he produced vivid pictures, stunning portrayals of light and shadow. Richard saw doctors and psychologists in Europe and the United States, but no one could explain how it was possible. His family and close friends had their own theories. Laurence thought that drawing set Richard's spirit free. Mike thought it was amazing. Ted thought it was nothing short of a miracle.

During one of the early tours, Richard appeared as the featured artist at the 1979 Special Olympics in Brockport, New York, about a thirty-minute drive from Rochester. He prepared a piece, *Superman*, for the event. In the darkly colored picture, Superman flies through a night sky punctuated by searchlights; lumps of color hint at the emblematic *S* on his chest, but his body is distorted.

Richard could remember precisely when he completed his pictures, but he generally couldn't articulate their meaning, so the viewer is left to wonder: Is this Superman a superhero *in spite* of his mangled form? Or does his unusual body—misshapen at first glance—*help* him do what no one else can?

⏤

Richard was technically a savant; he had an extraordinary spike in artistic skill accompanied by disability.

His passion for his specialty was intense and long lasting. By his family's count—a number that is surely short because they didn't begin cataloging Richard's work until he was seventeen—Richard completed 2,453 pictures during his fifty-three years and sold more than 1,300. His brother believes his work was featured in more than a hundred exhibitions in North America and Europe. Margaret Thatcher and Pope John Paul II both owned Richard Wawro originals.

Such powerful interests have long been associated with autism. Leo Kanner and Hans Asperger both remarked upon this characteristic in their first descriptions of the condition. One of Kanner's patients, Alfred L., for example, had "a marked tendency toward" such obsessions. As Alfred's mother put it, "He talks of little else while the interest exists, he frets when he is not able to indulge in it (by seeing it, coming in contact with it, drawing pictures of it), and it is difficult to get his attention because of his preoccupation." Asperger, too, noted a tendency toward strong, relatively narrow interests among his autistic patients, including a child enraptured with numbers (he loved complicated calculations), a child fascinated by complex machinery (his persistent questions were "impossible to fend off"), and another child enamored with chemistry (he used all his money to fund his experiments).

Today, such engrossing interests are one of autism's hallmarks. The *DSM-5* lists "highly restricted, fixated interests that are abnormal in intensity or focus" as one of the condition's core symptoms. Many autists demonstrate such circumscribed interests, and trains, physics, video games, and numbers are among the most common.

It's a startlingly similar sort of obsession that propels child prodigies to relentlessly pursue their crafts. As a child, Lauren Voiers colored on the wallpaper, drew on the carpet, and carved into woodwork; as a teenager, she stayed up late into the night, painting. Greg Grossman requested specialty food items as gifts, wrote about food for school projects, and dragged his mom to food trade shows. He cooked at home, he cooked at friends' houses; he hounded the chefs in his school cafeteria. William calculated every chance he got: he calculated dates and ages, swapped math problems with his uncle, and wrote out the powers of two for fun.

Several researchers have observed that autists are highly motivated to pursue their obsessions and that they view time so spent positively; they find their chosen activity fulfilling, and it makes them happy. This, too, is a characteristic shared with the prodigies. Each of them, beginning with Garrett James, who, even at four years old, took the stage with complete confidence and contentment, has a similarly positive take on his or her field. For them, playing the piano, painting, or engaging with complex equations isn't drudgery; it lights them up.

There are differences between the interests of the autists and those of the prodigies. While the prodigies tend to excel in culturally recognized domains—areas like music, art, and math—the autists' interests can lie in any area, ranging from traditional fields like music and math to more esoteric subjects like Pokémon and models of lawn sprinklers. From the autist's family's perspective, the choice of specialty may be no small matter; a dedication to music has a more straightforward vocational trajectory than a passion for Pokémon. Autists' families, just like prodigies' families, often appreciate their children's interests, though some also note that these interests can be disruptive to family life.

But in terms of what lies beneath those interests, and the extraordinary ability of prodigies and autists to single-mindedly pursue their passions, the division between laudable focus and unhealthy obsession seems a thin, fragile line.

⟶

By the end of 2011, Joanne had investigated nine prodigies. It wasn't much of a research sample by typical scientific standards, but for a prodigy study it was a strikingly large group. At that point, Joanne took stock of what she had seen so far.

Every prodigy she had met had an astounding working memory— a trait the prodigies share with autistic savants. Each of the prodigies had an unquenchable passion for his or her field, a trait similar to autists' tendency toward obsession. But those traits were only the beginning of the connection between prodigy and autism.

Autism is highly prevalent in the prodigies' families. Five of the first nine prodigies Joanne worked with had at least one close family member with autism. Three of the families combined to have eleven close autistic relatives, and one prodigy had five family members with an autism spectrum disorder. When these children look around at their relatives, it can seem that autism is everywhere.

Autism is more common among men than women, with boys landing on the spectrum about four times as often as girls. The reasons for this asymmetry are a bit murky, but some recent studies have investigated the possibility that higher levels of exposure to testosterone during pregnancy leave boys more vulnerable to developing autism. According to this theory, the elevation in testosterone leads to a dip in social skills, language development, and empathy but a spike in systemizing—traits that align with the symptoms of autism. The leading autism researcher Simon Baron-Cohen has thus suggested that autism is a manifestation of the "extreme male brain." Others have proposed that something *protects* females from autism and that females require a higher number of autism-linked genetic mutations to develop autism. Still others suggest that autism may look different in females and so females end up underdiagnosed.

Though her sample was small, Joanne saw this same gender skew in her initial population of prodigies: with seven boys and two girls, the breakdown was 3.5 boys to every girl, a ratio nearly identical to that for autism. It was another clue that underneath their surface differences, the two might have common biological roots.

Attention to detail is another link between prodigy and autism. This characteristic, the ability to notice and remember small things that others ignore or forget, has been described as "a universal feature of the autistic brain." Heightened attention to what the non-autistic brain dismisses as minutiae is thought to help those with autism excel at detail-oriented tasks, like identifying items or figures hidden in a larger design.

In one highly publicized 1996 incident, army rangers, Green Berets, marines, sheriff's deputies, and a host of other volunteers spent four days scouring the murky waters of a Florida swamp, searching for Taylor Touchstone, an autistic ten-year-old boy who had ventured out of sight while swimming. So treacherous was the swamp into which the boy had nonchalantly ventured that it had claimed the lives of four army rangers the previous year. On the fourth day of the search, a man fishing for bass found Taylor floating naked in a river, bloody and scratched but otherwise unharmed, fourteen miles from the spot where he disappeared.

Taylor's family believed his autism saved his life. Because Taylor often focused on tiny details, such as the knot in his bathing suit string, he traveled without panicking through a swamp teeming with alligators and poisonous snakes.

Autistic savants share this trait and sometimes put it to spectacular use. The artist Stephen Wiltshire, for example, has astounding attention to detail. When he was eleven, Stephen visited London's St. Pancras railway station, an intricate structure dripping with dormers, windows, and balconies, during a school trip. Afterward, Stephen reproduced it with startling accuracy. On other occasions, he has produced massive, astonishingly detailed, "frighteningly right" cityscapes after a short helicopter ride over a city.

Extreme attention to detail is also mirrored in the prodigies. The pilot study that Joanne conducted after meeting Garrett James revealed the prodigies' family members' elevated attention to detail. She administered the AQ (the same test she had used in that pilot study) to eight of the nine prodigies in her initial sample and found that the prodigies, too, demonstrated excellent attention to detail. They outscored those without any sort of autism diagnosis and even exhibited a greater attention to detail than those with Asperger's disorder. Jonathan Russell, for example, is so attuned to sound that he notices the pitch of car horns; when he rides the New York subway, he finds it jarring when the chimes on the cars are a half step off.

The prodigies' lives were rife with tantalizing hints that their precocious abilities were somehow linked to autism. If anyone embodied that increasingly robust connection, it was the ninth prodigy Joanne met: Jacob Barnett.

Chapter 6

Chromosome 1

Kristine Barnett's first pregnancy was a nightmare. The petite, fresh-faced blonde, then just twenty-four, developed preeclampsia, a pregnancy complication associated with high blood pressure. Her bouts of preterm labor forced her into bed rest. She took multiple emergency trips to the hospital. Three weeks before her due date, Kristine went into labor again; this time, her son, Jacob, was born.

Kristine and her husband, Michael, a quick-witted man who speaks in rapid bursts of words, brought Jacob back to their home in suburban Indianapolis. Less than a week after giving birth, Kristine returned to work at her in-home day care. She brought Jacob along with her.

With Jacob almost always by her side, Kristine had a front-row seat for his "firsts," a series of milestones that ripped past her. He said his first word, "ragù," at just three months old while Kristine was making spaghetti. Once his lips formed the word, he repeated it over and over again. "He was going absolutely crazy with spaghetti sauce," Kristine said. "I think my reaction was, he's really into this spaghetti sauce. Everybody look at this, this is insane. This kid really likes spaghetti."

He loved puzzles. Before he was a year old, he would take a dozen or so of the Peg-Board puzzles Kristine had at the day care—animal puzzles, truck puzzles, any other type of puzzle he could get his hands on—and dump all the pieces into a pile. He would pick up a piece and sift through all the puzzles until he found its place. He always knew which puzzle to look for; he always oriented the piece

99

so that it slid right into place. He would keep going until he finished all twelve puzzles.

Kristine and Michael got a glimpse of Jacob's memory during a trip to a furniture store. Michael and Jacob hung out in a waiting room stocked with books and toys while Kristine picked out a couch. Jacob gravitated toward the puzzles and plowed through one with alphabet letters. When Kristine returned, he recited the alphabet forward and backward. Kristine assumed he was just reading the letters in front of him. But even after they left the store, he kept saying the alphabet—forward and backward.

Around the time Jacob was six months old, he began memorizing shows, Baby Einstein videos, and movies—whatever he saw on TV. When Kristine was in the kitchen, Jacob provided running commentary from cooking programs. Dice this, he would say, or sauté that; chop those carrots.

At almost a year, Jacob homed in on a computer in the day care. He fished CD-ROMs out of a small basket, and using the arm of the couch to pull himself up, he inserted them into the computer. The programs recited books like *The Cat in the Hat* and *Horton Hears a Who!* while displaying the words on the screen. Jacob watched attentively. He burned through all the CDs at the day care. "He just went nuts with it," Kristine recalled. "Nobody else could touch it."

Jacob wanted—almost needed—to immerse himself in the worlds the books created. He demanded toy versions of the characters from *The Cat in the Hat*; he insisted on eating green eggs and ham; after he read *One Fish Two Fish Red Fish Blue Fish,* the family had to get a goldfish.

By the time Jacob was one, he could read short words. Not sound them out—*read*. He had spent so much time with his books and CD-ROMs, he knew the words by sight.

Jacob looked like a typical toddler. He had an inquisitive face, round cheeks, and light brown hair. He did typical toddler stuff, too.

He loved stuffed animals; he wrestled with his father. The day-care kids dressed him up and played patty-cake with him.

But even though he was actually far from typical, his family took even his most advanced abilities in stride, at least most of the time. Kristine had also spoken early, and by ten months old she could identify every plant in the local garden shop—the lipstick plants, the spider plants, the philodendrons. "When Jake came along, he did crazy stuff, but Kristine had done crazy stuff before," Kristine's mother, Becky Pearson, said of her daughter's and grandson's rapid development. "So we just kind of took him for what he was."

Around the time Jacob was fourteen months old, there were changes—small ones at first. He developed chronic ear infections. He didn't want to wrestle with Michael anymore. The sound of his talking and laughing, joyful noises that had once filled the house, became rare.

His early interest in letters and a similar interest in light, once charming, grew insatiable. They began to pull at him, sucking him into an isolating world. He carried his alphabet magnets with him everywhere. He stared at light on the wall, light on the water, and light on an apple; he used his hand to make shadows for hours at a time. At the day care, Jacob stopped chasing the other kids. He took a book or a stuffed animal and a flashlight and climbed into tightly enclosed spaces. If Kristine tried to pull him out, he cried.

A specialist confirmed that Jacob had significant developmental delays. He started seeing a speech therapist and then a developmental therapist. But instead of getting better, Jacob withdrew further. He stopped saying good night; he stopped saying "Mommy" and "Daddy." Soon the child who had said his first word at just three months old could only repeat memorized phrases like a lyric from a song or something he had heard someone else say. Physical contact repulsed him. If Kristine put him on her lap facing her, he would flip over and strain away from her to avoid eye contact.

Jacob was examined again a few months after he turned two. He recited the alphabet forward and backward and zipped through a puzzle, but he didn't respond to the evaluator or make eye contact. He refused to point at a circle, stack rings, or sing.

Kristine's great fear was an autism diagnosis. At the time, her impression of autism was heavily influenced by Dustin Hoffman's portrayal of an autistic savant in *Rain Man*. Kristine couldn't see her son in the same category as the character from the movie.

Jacob was instead diagnosed with Asperger's disorder. At first, that term, which sounded distinct from autism, brought Kristine a great sense of relief. It took a while for her to accept that autism and Asperger's weren't wholly unrelated diagnoses.

It hardly mattered, though. Jacob withdrew further. He stopped speaking altogether. He refused to eat anything except plain foods; if there were no pancakes for breakfast, he melted down. Spinning objects fascinated Jacob, and he watched them with frightening intensity. He developed an obsession with his alphabet flash cards, and he carried them everywhere. He loved to stuff himself into small spaces; Kristine sometimes found him at the bottom of the closet, on a shelf in the armoire, in plastic storage tubs, or in a laundry basket. When Jacob was reevaluated just before his third birthday, his diagnosis was revised again, this time to full-blown moderate to severe autism.

From the professionals, the Barnetts received a clear message about next steps: if they wanted to help their son, they had to get in as much therapy as they could as quickly as possible. Jacob did speech therapy, occupational therapy, physical therapy, and developmental therapy. They tried out a traditional behavioral therapy program and then switched to a more child-driven approach. Jacob spent hours stacking rings on a stick or trying to drop a ball in a cup. He practiced facial expressions and sounds, worked on holding a pencil, and tried to take the stairs one leg at a time.

The therapy continued even when Kristine and Michael's second son, Wes, was born and began having health problems of his own.

Wes was having seizures and couldn't swallow fluids; he often seemed to be in pain. The family took multiple trips to the emergency room with him, and more than once they feared for his life. Still, Jacob had therapy before the day care opened, therapy during the day, and then more therapy at night. When Jacob went to sleep, Kristine and Michael researched autism therapies and support groups. Every night, Kristine got in the shower and cried.

Months went by without Jacob's speaking. "It didn't matter how much therapy we were doing. If there was a rainbow, he was losing colors every month," Kristine said. "It wasn't working."

Jacob was listless and bored during therapy, but afterward he did the things he enjoyed. He spun balls. He drew shapes. He strung yarn through the kitchen, creating colorful, intricate webs. He studied the angles of light and shadows. He could tell the time—precisely—by the shadows on the walls. He wasn't talking, but sometimes he communicated through light. If Kristine was sad, he would take a faceted glass, angle it so it caught the light, splash a rainbow across the room to comfort her, and then run away.

The summer Jacob was three, there was a break in his state-provided therapy. One of the programs ended on his third birthday; it wouldn't pick back up until the fall, when Jacob would start special-ed preschool.

At the time, a lapse in therapy was unthinkable. Michael and Kristine filled in themselves, putting their own twist on the activities. Kristine created a mesh sling that Jacob could climb in to achieve the smooshed feeling that made him comfortable; he did parts of his therapy from inside. She poured thousands of dried beans into a wooden play table and let Jacob burrow in them; from there, she coaxed him into working on social goals. Kristine tried PECS, the picture-based communication system Lucie had used with Alex, and it worked. Within a few weeks, Jacob could point to the cards.

One day that summer, after watching the day-care kids laugh and play in the sprinklers, Kristine realized that for all their innovations with therapy, many of which Jacob enjoyed, that was all it was—therapy. His day was packed with it. Her three-year-old son was missing out on the exuberance of childhood. He wasn't having any fun. "He's my son. He's this beautiful child, and he deserves a childhood as much as anybody else does," Kristine remembered thinking.

Kristine made a course correction. They kept at Jacob's therapy during the day, but in the evenings they backed off. At night, Kristine drove Jacob out into the countryside, and they played jazz music and danced. They looked at the stars and ate Popsicles. They made s'mores or drove out to her grandfather's land. During the day, between therapy sessions, Kristine dragged Jacob outside and blew dandelions at him or turned on the sprinkler.

One night, a few months after they began their nighttime adventures, Kristine tucked Jacob in. She said a few words to him, the same soothing mantra that she repeated every night: "Good night, my baby angel. You're my baby angel, and I love you." For the first time in over a year, Jacob hugged her back. Then he spoke: "Night-night, baby bagel." Even now, Kristine cries when she remembers the sound of his words after so many months of silence.

⌁

That fall, Jacob began special-ed preschool.

To Kristine, it never felt right. She had spent the summer trying to give Jacob a taste of carefree childhood, something closer to ice cream and swimming pools than to task repetition, and now he had to start school earlier than other kids. But school was where Jacob got his therapy, and he certainly seemed to need it. After repeating Kristine's good night words back to her, he had begun speaking occasionally, but he still couldn't engage in conversation. So every morning, Jacob got on the bus.

A couple of months into the school year, one of Jacob's teachers came to the Barnetts' house for a conference. She brought up the alphabet cards Jacob carried to school every day. The cards were special to Jacob:

He kept them in his pockets, pants, and shoes. He stashed them around the house and got upset if anyone touched them. To Jacob, the cards were more than a learning tool; they were dear friends. But the teacher was concerned that Kristine and Michael needed to adjust their expectations. Jacob's education was geared toward life skills: helping him learn to dress himself, tie his shoes, eat with a spoon, and stay in his chair. Learning the alphabet wasn't on the agenda. Neither was learning to read.

That night, after the teacher left, Kristine checked on Jacob. As usual, his alphabet cards were everywhere, spilled across his puppy-and-pickup-truck comforter. She put her hand on his back and thought about all the things she had seen him do before his autism diagnosis—she thought about him reciting the alphabet and studying *The Cat in the Hat* on the computer—activities he loved. "I thought, well, they just have all given up on you, Jacob, but I haven't," Kristine said.

The next day, the school bus came and went. Jacob didn't get on it. Kristine didn't intend for him ever to get on that bus again.

There was paperwork to withdraw Jacob from school. The administrators emphasized that they didn't recommend taking Jacob out of school. They urged Kristine to talk to a doctor. At least send him for his therapy, they pleaded. But Kristine refused to send him for even that portion of the day; she felt that his self-esteem was at stake. "I didn't want him to feel like everybody has given up on him. I didn't want him to feel that; I didn't want him to know that," Kristine said. "And he would know. He would *know*."

Once Kristine had Jacob home again, she had to figure out what to do with him. She decided to help him focus on his interests—even if she didn't understand them. As Kristine once said during a TV interview, "Some of these things that he liked to do were considered repetitive behaviors. So he would, you know, he would play with a glass and just look at the light, twisting it for hours on end, and instead of taking that away, I would give him 50 glasses filled all different levels and let him explore."

Jacob learned about chess and was soon beating adults. He studied shapes. He taught himself Braille. He was intrigued by planets and particularly fixated on Pluto's distance from the sun. During trips to the library, Jacob sought out books about space. If those books represented Pluto as closer to the sun than it should be, he would rip the picture of Pluto out of the book and move it farther away.

On a trip to Barnes & Noble, a three-year-old Jacob discovered an astronomy textbook on the floor. He looked through it, drinking in the minuscule text and the maps of the solar system for more than an hour. He refused to leave without it. The book seemed far too advanced for him, but it was on sale, and Kristine bought it. Jacob couldn't get enough of it. He brought it with him everywhere, dragging the large tome around by its cover.

When the Butler University planetarium put on a special program on Mars, Kristine took Jacob to see it. During the presentation, the lecturer asked if anyone knew why the moons around Mars are elliptical. Kristine was shocked when Jacob responded with a question about the size of the moons and then answered the lecturer's original question: he said that because the moons were small with a small mass, their gravity was too weak to round them into spheres. As Kristine remembers, it was more conversation than she had ever heard from Jacob. "Then I knew, okay, he doesn't just know the information; he's able to understand the information, and he really knows how it all works together and what it *means*," Kristine said, "on this really crazy, crazy level for a three-year-old."

Jacob's interests engaged him in a way that nothing else did. The more time Jacob spent learning about astronomy, the more he interacted with those around him. Diving deep into one of his powerful, narrow interests—a classic autism symptom—*eased* his isolation. Kristine and Michael let him dive in as deeply as he wanted.

⌒

The fall he was five, Jacob enrolled in a mainstream kindergarten class.

He and Kristine had prepped for it like a stealth military operation. Kristine wasn't concerned about academics. Jacob could read; he could calculate. His abilities on kindergarten basics were almost absurdly advanced. It was the social aspect of kindergarten—the group play, the communication, the ability to follow directions—*that* was the element that she worried could bring it all crashing down.

They practiced every night. Sometimes they prepped on their own; sometimes they prepped at Little Light, a twice-weekly "kindergarten boot camp" Kristine had organized for kids with autism. The program had two goals: to help the kids develop their passions and to give them the tools to make it through circle time.

Jacob fought for every inch of progress. He practiced sitting still; he practiced engaging in group activities. It took nearly a year for Jacob to master sitting next to another child for ten minutes, but he learned to do it. Kindergarten would be the test of whether he had learned it well enough.

When school started, there were bumps: Jacob struggled with social activities and communication. He had a hard time when things were out of the ordinary, like when the school had a pajama day or his class had a substitute teacher. But for the most part, he did fine. He read the class books on the weather and rocks and tried not to let on about how advanced his academic abilities were. He made it through kindergarten.

The next few years felt like steady progress. It was still difficult to have a conversation with Jacob about anything other than science, but he began connecting with some of the kids in the neighborhood over video games. Kristine set up a sports program for kids with autism; there, Jacob met Christopher, a child who became a close friend.

But as the social struggles eased somewhat, problems bubbled up on the academic front. Jacob's boredom with the curriculum was wearing on him. The Barnetts headed it off as best they could. They spent most Saturday afternoons at Barnes & Noble, where Jacob used his two-book allowance on reference books and science textbooks. He loved reading about history and memorized the presidents and a motley assortment of facts about their lives and terms. As Kristine recalls in her memoir, *The Spark,* when Jacob discovered the test prep section of the bookstore, "he looked at me reproachfully, as if I'd deliberately been withholding this wonderful treat." He particularly loved working his way through the math problems in the GED prep book.

But by third grade, the reference books weren't enough anymore. At home, he stayed up late reading. Kristine found him hiding in the bookshelf again, just as he had during the most isolated days of his autism. He didn't want to go to school. Jacob had been pleading with Kristine to help him learn algebra, so she hired her aunt to teach him. They perused the NASA Web site and watched *Cosmos* and videos about savants. It felt as if they were squeaking by, though, hanging on precariously to the progress they had made.

When the Butler planetarium, Jacob's favorite haunt, closed for the winter, Kristine panicked. "That was like the whole world to Jacob," Kristine said. "I was like, oh brother, this is gonna be really bad. He's gonna lose all his social skills. He's not gonna play with anybody. I have to have the planetarium." She called around, searching for another planetarium they could visit. Eventually, she wound up on the phone with a professor at IUPUI, the joint Indianapolis campus of Indiana University and Purdue University, who said that Jacob could sit in on his astronomy class.

Eight-year-old Jacob lit up. He was eager to attend. During class, he asked and answered questions. When the course ended, he informally audited another and then another. The classes brought something in him alive. He was fully, deeply engaged.

Word of the kid attending college classes got around. When Jacob was ten, the Barnetts got a call from someone at IUPUI who proposed that Jacob formally enroll in college through SPAN, a program geared toward allowing high school students to take college classes.

The idea sounded crazy. Jacob was still in elementary school. The Barnetts couldn't see a future in which Jacob never attended prom or went to a high school football game. Jacob wanted to do it, though.

He was accepted into SPAN and withdrew from elementary school. But after a debacle during his college welcome interview—Jacob spent much of it chasing coins that had fallen on the floor—IUPUI started him with just one course.

Jacob had time on his hands. Other kids might have played video games or lost hours watching TV, but Jacob dove into his own research on an expanded theory of relativity. During this unstructured time, a dam broke. Ideas burst from Jacob's brain like an unstoppable torrent of water. They flowed constantly—when he was working, certainly, but also when he was at the playground and when he was having dinner. He once stabbed his whiteboard with a fork because, in his haste to write down an idea that came to him while eating, he forgot to swap his utensil for a marker. He stopped sleeping.

The Barnetts grew concerned. They took Jacob to see his pediatrician and enrolled him in a sleep study. But nothing was *wrong*. Jacob was just completely engrossed in physics.

Kristine was still eager to get Jacob out of the house and away from his work, though. She thought an expert in the field might help him move forward on whatever problem he was working on and hopefully ease its grip on his mind. She contacted an astrophysicist at the Institute for Advanced Study in Princeton, New Jersey, who confirmed that Jacob was working on some of the toughest problems in astrophysics and theoretical physics.

After another semester of SPAN, Jacob applied for college through the traditional channels. He was accepted and awarded a scholarship. Formally enrolling in college brought new challenges: Jacob was too small to carry all his textbooks in his backpack; Kristine worried about him on campus. But Jacob found a home in the Honors College, a program with a suite of rooms in the library where he could hang out between classes. He tutored other students in math and science, and many of them began to treat him like a younger brother. To twelve-year-old Jacob, college felt like home.

In Jacob, the connection between autism and prodigy is palpable. As a kid, he *had* autism (which raises the interesting question of whether he should technically be considered a prodigy or a savant—and whether that distinction is really as clear-cut as the terminology makes it sound). He had the communication difficulties; he had the social difficulties.

He also had a tendency toward obsession (or a tremendous passion for his interests, depending on how you look at it). When Jacob thinks back on the most isolating days of his autism, he remembers already being highly focused on the interests that would later catapult him to college. "I was focusing on things in such extreme detail that it seemed like I wasn't thinking at all," he said during his TEDx-Teen talk, "Forget What You Know."

Even after Jacob no longer had social and communication difficulties, his obsession with math and physics persisted. He stayed up late into the night devouring the subjects; taught himself much of geometry, algebra, algebra II, trigonometry, and calculus just so he could sit through a calculus review course; and created a home laboratory equipped with whiteboards, the periodic table, space posters, and an oscilloscope (an instrument that measures voltage changes over time).

His memory was breathtaking—just like that of prodigies, just like that of savants. Once, when a three-year-old Jacob was shopping with his mom and Wes, he listened to the songs played by a series of music boxes. He found a keyboard in the store and played the songs from memory. Around the time he was four, he memorized a U.S. atlas, and his family nicknamed him JPS—Jacob Positioning System—for his ability to navigate even unfamiliar cities.

He had the extraordinary attention to detail. He picked up on even seemingly inconsequential facts, like the number of blue cars in a parking lot and what percentage of the electoral vote President James

Buchanan received. When Jacob was four, his grandmother gave him a large, detailed map of Indianapolis. He studied it for a few minutes and then told his grandmother the map was out of date. A small section of I-465 had been renamed I-865, but the map didn't reflect the change. It was inaccurate; Jacob didn't want it.

Jacob embodied nearly all of these connections between autism and prodigy; he also possessed another, less well-known trait: synesthesia.

It sounds more like the stuff of science fiction than serious academic research. Synesthesia occurs when a certain type of stimulus, maybe a musical note or a number or a word, elicits a response, such as a color or a taste, usually associated with a totally different type of stimulus. Those with synesthesia don't just see what others see or hear what others hear or taste what others taste; they see and hear and taste *more*.

Synesthesia comes in a variety of forms. Sometimes numbers or words evoke a particular taste (like when the word "jail" calls to mind cold bacon) or months occupy a spatial location (like December sitting an arm's length away from the left of the body). Some people personify letters, numbers, or months. One woman perceives August as a chubby boy prone to becoming defensive and the number 8 as entangled in a love triangle—she is dating 9 but loves 7.

The most common form of synesthesia involves the pairing of letters, numbers, or words with a particular color. Researchers have picked up on some trends in this area; those who associate letters with colors frequently perceive *a* as red, *d* as brown, *o* as white, *r* as red, and *y* as yellow. Some are tougher to pin down; the unreliable *p* was primarily viewed as green in one study, pink in another, and blue in a third.

These unexpected responses to letters, sounds, numbers, or other stimuli are automatic. They also tend to be constant over time. Tests taken months or even years apart reveal a consistency rate above 90 percent; if an individual with synesthesia perceives the letter *s* as lazy, self-centered, and feuding with the letter *p*, *s* will never lead an industrious life or think of others, and its quarrel with *p* will continue forever.

Synesthesia is particularly interesting in the context of prodigies and autism because it comes with a memory boost: Synesthetes, as those with synesthesia are known, have enhanced recall for the types of stimuli that induce their synesthesia. Those who perceive letters as having a color, for example, have a superior ability to recall letters, while those who perceive time as having a particular spatial location have exceptional recall for dates.

Jacob's synesthesia memory boost is associated with numbers. When Jacob thinks about a number (say, 3), he doesn't just picture the numeral; he perceives it as having a specific color (like red) and a specific shape (like a triangle). As Jacob once put it during a conversation with a reporter, "Every number or math problem I ever hear, I have permanently remembered." But he has trouble remembering smells and conversations.

Over the years, there have been several anecdotal reports of an association between autism and synesthesia, including a case study of an extraordinary British savant, Daniel Tammet, the man who recited 22,514 digits of pi from memory. Daniel can also conduct complex calculations in his head quickly and without error. He speaks ten languages and learned Spanish over the course of a weekend.

Like Jacob, Daniel has synesthesia. He sees numbers as having a shape, color, and texture and perceives some words as having a color. He also has Asperger's disorder. It's no coincidence that Daniel has both conditions; a recent study found that adults with autism spectrum disorders are more than three times more likely to report synesthesia than non-autistic adults, making synesthesia another link between autism and talent and potentially another link between autism and child prodigies.

⟶

In 2011, a reporter from a small Indiana newspaper wrote a story about Jacob. Two months later, the *Indianapolis Star* published a lengthy profile on the twelve-year-old scientist who was trying to disprove the big bang theory, and that story got picked up by a wire service. Word of the whiz kid was rehashed in print and plastered all over the Internet. The full weight of the media crashed down on the Barnett household.

It was an exciting, frightening, sometimes overwhelming time. It was also eye-opening. As Kristine told Glenn Beck later that month, she hadn't realized that Jacob, who was sitting next to her in a backward baseball cap and a pi T-shirt, was *that* unusual. "I really just thought he was, you know, just another smart kid," she said.

It was in the midst of this media frenzy that Joanne contacted the Barnetts about her research. Kristine was skeptical. "At first I sort of thought, well, I don't know about that," she recalled. But then Joanne asked if Jacob might like to go to Cedar Point, a Sandusky, Ohio, amusement park jammed with roller coasters. The Barnetts packed their kids into the car and began the five-hour drive to Sandusky, eager to talk to someone who might provide a new perspective on the child who couldn't get enough theoretical physics.

The Barnetts also consented to one more interview. They had been approached by *60 Minutes,* and convinced that the reporters and producers there would do a thoughtful piece, the Barnetts said yes. They pointed them to Joanne as a prodigy expert.

Joanne arrived at the family's hotel toting a large poster of the United States that she had bought from a teacher-supply store. With the cameras rolling, Joanne showed Jacob a series of states and then asked him to repeat them back to her. Jacob zipped through the list of twenty-eight states, both forward and backward.

After *60 Minutes* finished taping, Joanne administered the more serious tests. Jacob issued a knockout punch in the working memory test, just as the other prodigies had done. Afterward, he was chatty and eager to explain the fourth dimension.

When the segment aired the following January, it opened with a picture of Jacob, his baseball cap slung backward over his head, his freckled face happy and curious. The first few minutes focused on Jacob's accomplishments, his autism diagnosis, and his love for math and physics. Jacob used a light box to demonstrate the way he saw numbers: as colored shapes, often layered on top of one another. As the segment continued, it shifted to Joanne's work with Jacob. The correspondent Morley Safer explained the memory test Joanne had done with Jacob for the cameras. When Morley met with Jacob three months later, Jacob was still able to recite all twenty-eight states, in order, backward and forward.

At one point, Morley asked Jacob about his autism, noting that his parents said that he was proud of the condition. Just as Joanne had long suspected, Jacob believed that he had succeeded not *in spite* of his autism but *because* of it. "That, I believe, is the reason why I am in college and I am so successful," Jacob said. "It is the rise as to my love for math and science and astronomy and it's the reason why I care. Otherwise, I wouldn't have gotten this far." In Jacob Barnett, the connection between prodigy and autism wasn't just tangible; it was celebrated.

⌒

The prodigies and autists had behavioral similarities; they had cognitive similarities. But the biggest question remained: Was the connection between the two genetic?

Joanne partnered with two Ohio State genetics researchers, Chris Bartlett and Stephen Petrill, to find out. After Joanne appeared on *60 Minutes*, she spent a spring break zigzagging back and forth between prodigies, asking the kids and any willing family members (some autistic, some not) to spit into small vials.

Back in Columbus, Bartlett and his team extracted the DNA from the saliva samples and prepared it for analysis. They then sent the DNA to an outside lab for genotyping. That outside lab extracted each individual's genetic code from his or her DNA and returned the raw data, "a big text file with a lot of numbers," as Bartlett put it, to the Ohio State team.

The idea was to find out whether autists and prodigies—two groups of people who, from the outside, often look completely different—shared genes that explained their commonalities. Once you stripped away their outer trappings, did the two conditions have common genetic roots?

The chances of finding anything were slim. The team used a linkage analysis, a study design that assumes the sought-after genetic mutation has a big impact. It was unlikely to pick up on anything with a relatively small effect. And they were working with a very small sample. The team had DNA from eleven prodigies and their families, but only five of those families included a sibling, a prerequisite for inclusion in the initial analysis. The mutation they were looking for would have to be a powerful one (highly penetrant, in genetics speak) to show up in an analysis on such a small group.

The team coded both the prodigies and the autists in the sample as "affected" and all the non-prodigious, non-autistic relatives as "unaffected"—essentially pretending that the prodigies and the autists were equivalent to each other. The idea, then, was to look for a slice of DNA inherited by the prodigies and the autists but *not* by their non-autistic, non-prodigious siblings.

The team plugged their equations into the computers and waited. Some of the analyses took days; one left the computer churning for nearly a week. The main results, when translated into a graph, look like an EKG. A single dark line hovers around the baseline, and then occasionally darts upward, indicating a potentially relevant place on the genome.

There were several blips on the prodigy-autism radar—something interesting on chromosome 8, something interesting on chromosome 20, maybe something worth looking into with a larger sample on chromosomes 5 and 11.

There was also one clear hit. At the very beginning of the chart, near the middle of chromosome 1, the line leaped up. On the short arm of chromosome 1, a location known as 1p31-q21, the team found something. They couldn't pinpoint the precise genes at play, but DNA in this region seemed tied to both prodigy and autism.

The researchers conducted two statistical analyses on the data; in both cases, the finding on chromosome 1 was statistically significant. They pressure tested the result. Would the link be cleaner and clearer if they changed their assumptions? What if they assumed that this particular region was tied only to prodigy? What if they assumed it was tied only to autism? Nothing came close. Not only did no other model achieve statistical significance, but the next-best fit with their data, the model that assumed that prodigy and autism did *not* have common genetic roots, was *fifty times* less likely to be true than the model that assumed that they did. "We did gamble. We put it all on the line by coming up with what we thought were reasonable tests of this hypothesis, and we failed to falsify it," Bartlett said. "In terms of science moving forward, this is really as good as it's going to get."

In at least some of the families, there seemed to be a genetic link between prodigy and autism. Despite their outward differences, the two groups had a common genetic core.

⟶

The link isn't entirely clean. There are a few family members—"carriers," of a sort—who are neither prodigious nor autistic but who seem to have a mutation in the same location on chromosome 1.

It's unlikely that those family members are secret prodigies or autists; both conditions are typically pretty hard to miss. The more likely explanation, the team believes, is that the identified region on chromosome 1 doesn't act as an on-off switch for prodigy and autism; both conditions are far more complex than that.

Perhaps this area on chromosome 1 is tied to a particular characteristic shared by prodigies and autists, like exceptional memory, a sharp eye for detail, or a tendency toward developing passionate interests. According to this theory, those non-prodigious, non-autistic family members who share the variation on chromosome 1 would possess that particular trait but lack whatever genes are responsible for the other behaviors of autism, the other behaviors of prodigy.

But which trait could it be? From the team's data, there's no way to be sure—or even to know if this theory is correct. And chromosome 1p31-q21 isn't exactly a hot spot in the search for autism's genetic roots. The short arm of chromosome 1 has come up a few other times in autism genome scans. In each instance, the region generated enough friction that it seemed worth mentioning, but in each case it fell short of statistical significance. Perhaps this is partially because there's a rare variant at play in these families; perhaps it's because there aren't many people tearing through the genome searching for the genetic source of the *strengths* associated with autism.

In either case, it's a preliminary result based on a small sample. It needs to be replicated; the team needs to conduct additional analyses to isolate the specific mutation (or mutations) at play. But if the preliminary results hold up, then the team has uncovered a major piece of the prodigy puzzle. The results would mean that prodigies and autists don't just have shared behaviors and cognitive traits; their similarities extend all the way down to their genes.

Chapter 7

The Empathy Puzzle

In the late 1970s, Uta Frith encountered someone unexpected: an autistic boy who spoke with remarkable fluency.

He was something of a puzzle. Frith had already been studying autism for over a decade; while some of the boy's behaviors certainly appeared autistic, his speech cut against her beliefs about autists' abilities. She and a colleague began an earnest conversation about whether such a highly verbal child could actually have autism.

The prevailing picture of autism at the time was clearly very different from the one we have now. As Frith recalls, researchers thought that autism was extremely rare, affecting perhaps four people in ten thousand; she and her colleagues thought that they knew of nearly all the autists living in London. The autists they knew, moreover, were all children, and almost all of them were boys.

Autism was also thought to be closely linked with intellectual disability. While today there are prominent figures such as Temple Grandin who are highly intelligent, professionally accomplished, *and* on the autism spectrum, in the 1970s and 1980s it was widely believed that the vast majority of autists—some studies put the figure as high as 94 percent—had IQs in what was then called the mentally retarded range. The *DSM-IV*, published in 1994, claimed that "most cases" of autism had "an associated diagnosis of mental retardation."

This relatively narrow perception of autism was initially reinforced (or perhaps reflected) by a particular turn in autism research: the quest to uncover a cognitive explanation for autism. It's an effort that began in the 1980s when a number of autism researchers dedicated themselves to identifying a cognitive abnormality—some oddity in the way the brain processed information—that could explain all of autism's core symptoms.

Those core symptoms were all deficits, such as social and communication difficulties. Most assumed that these *behavioral deficits* had to stem from a *cognitive deficit*. As a result, the first scientists to partake in the hunt for a cognitive explanation for autism looked for mental weaknesses rather than strengths and perceived those weaknesses as autism's defining characteristics.

It's not surprising then that when these researchers began generating cognitive theories of autism, those theories were almost entirely deficit focused. The executive function theory, for example, was based on the idea that autists had an impairment in those abilities related to setting and working toward goals, such as planning and impulse control, that resulted in their insistence on sameness and repetitive behaviors. According to the weak central coherence theory, autists had an imbalance in the way they integrated information—basically trouble seeing the forest for the trees—that resulted in their communication and social difficulties and repetitive behaviors, as well as a "peculiar pattern of intellectual abilities."

Autists, as portrayed by these theories, appear almost completely divorced from prodigies. The same is true of the mind blindness theory of autism, a cognitive theory that suggested that autists lacked the ability to empathize—a trait that prodigies have in abundance.

⌐

It all started with a chimpanzee.

In the 1970s, two researchers set out to test whether a chimp named Sarah had a "theory of mind"—an understanding that others had motives and beliefs different from her own. They showed Sarah a series of videotapes in which an individual struggled to reach bananas, escape from a cage, or complete other tasks. They found that Sarah could use photographs to indicate the actions needed to reach that goal and that she systematically chose different outcomes for different actors. The researchers argued that Sarah could thus infer the intentions and knowledge of others, demonstrating that she had a theory of mind.

It was an intriguing concept, and scientists began examining theory of mind in humans. In the early 1980s, two researchers concluded that most children developed theory of mind between the ages of four and six.

But was this true of all children?

One team of researchers thought not. Frith and Alan Leslie, both then psychologists at the MRC Cognitive Development Unit in London, and Simon Baron-Cohen, then a doctoral student, predicted that autistic children might not have such an ability.

To find out, Baron-Cohen put sixty-one children through what would become known as the Sally-Anne test. The child-subject looked on as a doll, Sally, placed a marble in a basket and then left the area. In the Sally doll's absence, the Anne doll removed the marble from the basket and placed it in a box. The Sally doll then returned, and the researcher asked the child-subject where Sally would look for the marble.

Most typically developing children passed the Sally-Anne test. The children said that Sally would look for the marble in the basket where she left it, demonstrating an understanding that what they knew (the marble was in the box, where Anne placed it) was different from what Sally believed (the marble was in the basket, where Sally left it). But most autistic children said that Sally would look for the marble in the box, where the child-subject knew that it actually was, even though the marble had been moved in Sally's absence. The researchers concluded that perhaps autistic children couldn't differentiate Sally's knowledge from their own, a lack of theory of mind that Baron-Cohen (the team member who became most closely identified with this line of research) eventually described as mind blindness.

It was a firecracker of a study. Theory of mind research proliferated as psychologists examined autists' abilities to detect faux pas, discern a person's feelings from photographs of his or her eyes, and respond to the emotions of a videotaped child. In nearly every case, they found some evidence that autists lacked typical theory of mind.

It wasn't long before scientists linked a lack of theory of mind to a lack of empathy. A few autism researchers had noted in the past that autists seemed to lack empathy, but as researchers investigated theory of mind, this notion became firmly ingrained in the popular consciousness. Baron-Cohen at one point described mind blindness as "deficits in the normal process of empathy." He and his colleagues at the Autism Research Centre developed the Empathy Quotient, a test meant to measure empathy in adults. In piloting this test, the researchers predicted—and found—that those with high-functioning autism or Asperger's disorder scored significantly lower on the Empathy Quotient than those without autism. A group of researchers later characterized a lack of empathy as "one of the key characteristics" of Asperger's disorder; others described autism as "marked by empathy deficits."

The perception that autists lack empathy made prodigy and autism appear completely divorced from each other; after all, for the prodigies empathy seems to be second nature.

⟿

Jourdan Urbach is of medium height and build with thick brown hair, thick eyebrows, and a close-trimmed beard. He speaks with matter-of-fact authority about subjects ranging from nerve remyelination to the ethics of international volunteer work. There's an intensity about him, a palpable energy that propels him forward; it seems that momentum must always be on his side.

These days, he's an entrepreneur. He's the co-founder and chief technology officer of Ocho, a social video app. But entrepreneur is only the latest of Jourdan's professional incarnations. By the time he was twenty-two years old, the Yale graduate had already had several other careers—scientist, musician, and philanthropist.

The first of these careers was music. For as long as Jourdan can remember, there has been music in his head—sometimes original compositions, sometimes other people's music—and usually in his fingers, too. These days, that music is in the background. But it wasn't always so.

As a toddler, Jourdan ambled about his Long Island home carrying a tape player equipped with classical music cassettes. His mother, Deborah, was a cantor and gave voice lessons; when her lessons ran late, Jourdan would gather his favorite stuffed animals, lie underneath his mother's piano, and listen.

During one lesson, Deborah asked her twenty-two-month-old son to identify the note her pupil had sung. Jourdan correctly answered, "A." Deborah quizzed Jourdan on several more notes. Each time, the toddler answered correctly. Not long after, he identified the notes of the sounds made by the teakettle and the vacuum cleaner.

Deborah brought Jourdan to see a musician friend at a local school. He watched Jourdan bang around on a few instruments and then confirmed that Jourdan had perfect pitch; Jourdan could immediately identify a musical note plucked out of thin air. The musician advised Deborah to get Jourdan started on an instrument right away.

At an orchestra concert, a two-and-a-half-year-old Jourdan picked out the violin as the instrument he wanted to play. His parents bought him one—"the tiniest violin you could imagine," as his mother later recalled during an interview with the *New York Daily News*. Jourdan breezed through the Suzuki volumes of classical music. By the time he was four, he had mastered them all. At six, he was winning regional music competitions. When Jourdan was seven, a renowned violin instructor took him on as her youngest pupil.

Jourdan's passion for science developed alongside his abilities as a musician, and it was the combination of the two that would turn him into a tiny philanthropic powerhouse. When he was in first grade, he began a research assignment at his local public library. The topic was wide open. Jourdan typed "neurosurgery" into a library computer; as he recalls, he thought the human mind was an intriguing, relatively unexplored frontier. A list of books popped up, and Jourdan closed his eyes and put his finger in the middle of the screen.

He landed on *Gifts of Time,* a book about Dr. Fred Epstein, a pediatric neurosurgeon known for lifesaving operations in seemingly hopeless cases. Jourdan was enthralled by Dr. Epstein's story. He wrote the doctor a letter explaining that he was seven years old, a devoted student of neuroscience, and a concert violinist; then he requested an interview. Dr. Epstein invited him to Beth Israel Medical Center in Manhattan.

Jourdan arrived at the hospital with a large black tape recorder and a long list of questions. He met with Dr. Epstein and a neurogeneticist. Jourdan grilled the doctors for more than two hours. Afterward, Dr. Epstein took Jourdan on a tour of the hospital's tenth floor. Behind the glass wall of each room was a different child. Some were Jourdan's age; many were suffering. In what would become a pattern in his life, this deeply personal experience served as a call to action that led Jourdan to take on challenges seemingly far beyond his years. Jourdan asked how he could help.

"I anticipated med school would be a breeze and I could probably drop out of elementary school and go do that instead," Jourdan recalled. "He told me it would be a solid twenty years, and at that point I said, 'That's no *bueno.* What can I do now?'" Dr. Epstein suggested that Jourdan find a way to help the patients through his music, and Jourdan and his friends began giving monthly performances for the children at Beth Israel.

Two years after his hospital performances began, Jourdan met Jason (a pseudonym), a thirteen-year-old who was frequently in the hospital with recalcitrant spine tumors. Jason was a concert pianist and hated not being able to practice at the hospital. Jourdan, who acutely understood Jason's need to play, set out to raise the money to buy a piano for the children in the hospital. He wrangled together a group of musicians and organized a concert fund-raiser at a high school auditorium.

"We packed it. I think the folks at the hospital that we had been working with for God knows how many performances at that point really came through for us and rallied a whole bunch of folks to come out of the city to Podunk Roslyn," Jourdan said. "I think the community rallied around it as well."

The event generated thousands of dollars—there was enough to buy a piano for the tenth floor; there was money left over to begin a children's surgery fund. The success of the event led Jourdan to found a nonprofit organization that used musical performances to raise funds for charities, particularly those that benefited children. "I saw the force multiplier that fund-raising could be and decided to pivot," Jourdan recalled. "I think I was nine at that point."

⌒

That same year, Jourdan began an internship at Cold Spring Harbor Laboratory, a famed genetics and molecular biology research institution on Long Island. He came to the lab every week or two and spent a few hours shadowing Eric Drier, a postdoc studying memory. "I *really* wanted to work there. I was obsessed with Watson and Crick and DNA. I was a total fanboy," Jourdan recalled.

Jourdan shadowed Eric through experiments. To keep up with what was going on at the lab, he taught himself high school biology, anatomy, physiology, and chemistry. Sometimes Eric gave Jourdan assignments to complete. At one point, the two investigated fruit flies and memory; as part of the experiment, the fruit flies were occasionally zapped with sixty volts of electricity. Jourdan felt bad for the flies and tested the grid out on himself.

Jourdan and Eric chatted while they worked. They covered everything from mutant fruit flies to synaptic plasticity to a science fiction book Jourdan had written. A persistent theme through all of their conversations was Jourdan's desire to have a positive impact on other people. "He would talk about helping others pretty regularly," Eric remembered. "I found it very heartwarming that he was concerned— very concerned—about helping out other people."

At eleven, Jourdan made his Lincoln Center debut, performing as a soloist with the Park Avenue Chamber Symphony. Much of his family came to watch, but one of Jourdan's cousins was missing from the audience. Jourdan soon learned that she had been diagnosed with rapidly progressive multiple sclerosis (MS).

Jourdan again responded with every resource he had. He gave concerts at Carnegie Hall in Manhattan and the Shubert in New Haven to benefit the National Multiple Sclerosis Society. He paired up with a faculty member at Stony Brook University School of Medicine and began studying myelin repair and other MS-related issues. At sixteen, he interned at a Harvard Medical School lab and worked on the genetics of multiple sclerosis. He eventually served as the keynote speaker at an MS event in Minneapolis and was featured at the Connecticut chapter of the National MS Society's Annual Meeting and Awards Ceremony.

Jourdan enrolled at Yale at seventeen. Within a few months of his arrival, he launched the International Coalition of College Philanthropists, an organization dedicated to supporting college students' fund-raising. "I wanted a platform that was totally clean for kids to be able to go out and fund raise against and then funnel into microfinance projects, really simple stuff. Now it's common; at the time it wasn't," Jourdan recalled.

The impetus for Jourdan's charitable work was almost always a personal encounter, but once he started fund-raising, he was intensely, thoroughly logical. The point was to generate as much money as possible and to get those funds to the organizations that would make the best use of the money. He kept the overhead for his own nonprofit at zero. By the time he was thirteen or fourteen, Jourdan was scrutinizing financial disclosure forms as he weighed various organizations interested in his benefit concerts. He grilled the organizations on how they planned to use the money they received; he has a pet peeve about charitable organizations with bloat.

He used the media, too, to his benefit. He was often written up in magazines and newspapers and appeared on TV—all exposure that Jourdan believes amplified his philanthropic reach. "Those were all leverage for me when I was trying to recruit orchestras to work with me at no cost and all these other crazy asks I would make on people. I embraced the media because it gave me a microphone," Jourdan said.

By the time Jourdan was nineteen, his nonprofit had raised $4.7 million. That money had provided twelve children with brain surgeries, a thousand cochlear implants for children who could not hear, and hundreds of thousands of dollars in aid for multiple sclerosis research and services; it helped fund pediatric clinics in El Salvador and Ghana and a music therapy program at the University of Michigan's C. S. Mott Children's Hospital.

Jourdan was named one of the top ten youth volunteers in the nation by the Prudential Spirit of Community Awards, the *New York Post* awarded him a Young Heart Liberty Medal, and he received a National Caring Award and a World of Children Youth Award. *Teen People* named him one of Twenty Teens Who Will Change the World.

As a spokeswoman for the National Multiple Sclerosis Society once put it, Jourdan's work "dramatically illustrated that a small hinge can swing large doors."

⟶

Jourdan isn't the only prodigy with a mind oriented toward helping those around him. Many prodigies seem to need to do good in the same way they need to paint, perform, or calculate.

Jacob Komar was typing computer commands at two and studying computer manuals at five. He soon began writing code and dismantling and rebuilding old computers. When Jacob was nine, he acquired a heap of old computers headed for the trash from his sister's school in Connecticut. He repaired the machines, and for about a year he refurbished roughly two computers each week. He coordinated with a local social services office to identify families who needed a computer and then delivered the machines personally.

Jacob's computer repair operation expanded. Eventually, he had more than two hundred computers waiting to be refurbished. They filled the Komars' garage; the family's cars had to be parked outside. Jacob enlisted help from friends, classmates, and strangers in his quest. People across the country who wanted to start similar programs contacted Jacob. He spoke with many of them and created a how-to manual to provide guidance for these groups. Jacob's nonprofit partnered with other organizations and won a $1 million grant from the National Science Foundation to teach junior high and high school students technology skills. He teamed up with the Cheshire Correctional Institution to teach prisoners how to refurbish computers in the prison repair shop.

In the meantime, he had been zipping through his own education. At sixteen, he graduated from the University of Hartford with a computer engineering degree. The following fall, he began a Ph.D. program at Brown University. A year after he completed his master's degree in electrical engineering, Jacob flew to Peru to set up a computer lab for the Sacred Valley Project, a nonprofit that helps educate girls in remote areas. A few months later, he flew back to help install solar panels in a rural community without electricity.

Other prodigies demonstrate the same extreme empathy. When visiting his father at the hospital, the child chef Greg Grossman was moved by the young cancer patients. He raised money to donate electronic games and movies to the children awaiting treatment. As his culinary reputation grew, he cooked for countless fund-raisers, hosted tasting tables at charity events, and regularly flew to Ohio to support Veggie U.

Lucie's son William was so troubled by the goal of dodgeball—to get other people "out"—that he always tried to get "out" himself at the beginning of the game to avoid doing something unkind to another child.

Many of the prodigies seize any opportunity to use their special skills for the benefit of others; when faced with the plight of someone in need, they take the task upon themselves. It never seems to occur to them to wait for an adult to help.

The gulf between autism—as portrayed by the mind blindness theory—and prodigy seems vast. But the empathy gap may not be as large as it seems. There's another, newer theory of autism—the intense world theory—far removed from the deficit-focused orientation of the first generation of cognitive theories. It's built from the premise that perhaps the autistic brain isn't *less* of anything. Perhaps it's *more*.

～

In 2002, Tania Barkat was surrounded by rat brains.

The project she was working on had been suggested by her adviser, Henry Markram, a famous neuroscientist (he would later coordinate the Human Brain Project, a hugely ambitious effort to build a working model of the brain) at the Swiss Federal Institute of Technology. Markram had begun reading about autism when his son was diagnosed with the condition. He had spent his career studying brain circuitry, and he wanted to investigate the autistic brain at the cellular level using an animal model— thus, Barkat's rats.

Barkat, who was then a Ph.D. candidate, exposed some of the rats prenatally to a chemical compound, valproic acid (VPA), that increases autism risk. She then carefully preserved and prepped rat brain slices so that she could stimulate the brain cells and measure the response. She wanted to compare the rats' brains and see if she could identify any abnormalities in the VPA-exposed rats' inhibitory cellular networks.

Barkat examined various layers of the brain; she looked at different classes of cells. But after two years on the project, everything in the VPA-exposed rats still looked normal. Markram thought they had exhausted the approach.

But Barkat wasn't ready to give up. One of her colleagues was studying excitatory cell networks, which gave her a new idea. Maybe the problem wasn't with the approach; maybe it was with the types of cells she had been targeting. She decided to switch from studying inhibitory networks to studying excitatory networks.

After Barkat changed tactics, she quickly began to see differences between the VPA-exposed rats and their typically developing counterparts. She and a group of fellow researchers set about trying to pinpoint exactly what those differences were.

It turned out that the brains of the VPA-exposed rats were "supercharged." Their neurons generated many more connections to other neurons—more than 50 percent more—than those in the control rats' brains. When stimulated, the brains of the VPA-exposed rats reacted nearly twice as strongly as the brains of the control rats. The long-term impact of that stimulation was also much greater for the VPA rats: while both VPA and control brains demonstrated increased reactions to later stimuli, the increase in amplitude for the VPA brains was more than twice as large as that of the controls.

Through studies of live rats, the team discovered that their VPA rats had some unusual behaviors, too. They had overblown levels of anxiety and fear: they learned what to fear more quickly, generalized that fear more broadly, and were slower to release that fear than the control rats.

To investigate the roots of these behaviors, the scientists examined the rats' amygdalae. The amygdala consists of two relatively small, oval-shaped sets of neurons embedded deep within the back of the brain. It's considered mission control for fear; scientists believe that it's where we store our memories of fear and process threatening situations. After discovering the VPA rats' fear behaviors, the researchers investigated whether their amygdalae, too, were supercharged.

The answer was yes. The VPA rat amygdalae were more responsive to stimulation, and the effects of that stimulation were longer lasting.

From these findings, the intense world theory of autism was born. Henry Markram, his wife and fellow neuroscientist Kamila Markram, and Barkat proposed that the autistic brain's hyperactivity, its ability to form numerous, strong connections, results in heightened perception, attention, memory, and emotionality. These tendencies, the theory goes, could explain all facets of autism: Autists' withdrawal and repetitive behaviors might stem from their extreme sensitivity to stimulation, which might make some environments painfully intense. Autists' excellent attention to detail might be a by-product of heightened perception. Savant skills and exceptional memory might flow from the brains' ability to change and make new connections quickly.

The intense world theory also has something intriguing to say about empathy. The scientists believe that there's no autism empathy deficit at all. According to this theory, autists are too perceptive; they feel *too deeply* for others—so much so that they become overwhelmed by their feelings and withdraw or avoid social interactions.

It's an intriguing theory built not around deficits but, in a sense, around strengths. It's yet to be proved, and not all scientists support it.

But for many families and autists, this theory—particularly the empathy piece—better aligns with their experiences. Certainly some people recognize a failure to read emotional cues in the autists they know, but others perceive their children as extraordinarily attuned to others and give them nicknames like "emotional barometer" and "mood ring" for their abilities to sense others' feelings.

Some autists have articulated the way that this sensitivity can make social interactions feel like an assault, just as predicted by the intense world theory. One man commented that to him other people seemed like "emotional tornadoes"; a woman felt as if others' emotions were punching her in the face. Such social experiences can leave those with autism feeling exposed or overcharged: one man felt as if his heart and soul were "like an exposed nerve to the world."

As a result, some autists report experiencing the predicted emotional overloads. One woman wrote about her tendency to "go into sensory lock down," ensconcing herself in the safety of her "bubble." A woman with Asperger's disorder was distraught for an entire weekend because she thought she had killed a butterfly. Positive experiences, on the other hand, could also reverberate deeply. As one individual put it, a non-autist would need a whole evening of hugs and reassurance to feel as cared for as he did after receiving a pat on the back and a smile from another autist.

There's still much work to be done on the intense world theory of autism. But seen from the perspective of this theory, many autists have astounding intellectual capabilities *and* are highly sensitive to the plight of others—just like the prodigies. If mind blindness seemed to cast prodigies and autists as polar opposites, the intense world theory casts them as close cousins.

⟶

It had been there all along.

The idea that autism was linked to various strengths has appeared in academic writings since Kanner identified autism as an independent condition in 1943. In that first paper, he described the autists' excellent memories for vocabulary, rhymes, and patterns; he said the children were all "unquestionably endowed with good cognitive potentialities." The following year, Asperger observed that some autists demonstrate "a high level of original thought and experience." A late 1970s study concluded that approximately 10 percent of autists possessed notable abilities in music, art, and other areas; since then, scientists have identified other autism-linked strengths, such as excellent attention to detail.

For decades, these strengths were mostly relegated to the background as researchers sought a cognitive explanation for autism. But over time, researchers adopted a broader understanding of what autism could look like, and there was a renewal of interest in those intriguing abilities and strengths that had appeared in even the earliest autism studies. The deficit-focused orientation toward autism began to give way.

Frith was partly responsible for this shift. Her own perception of autism was already evolving when she read Hans Asperger's 1944 paper during a seminar. Frith was struck by Asperger's description of autism, and she eventually decided that it should be made more widely available. Her 1991 translation of his work into English put autism in a startlingly different light for many people.

In this paper, Asperger notes the children's social difficulties and repetitive behaviors as well as their highly original use of language, distinct areas of special interest, and excellent logical and abstract thinking. He claimed that autism could affect individuals of *any* ability level and emphasized that autists of high ability had extraordinary potential. As long as they were "intellectually intact," Asperger thought that professional success, "usually in highly specialised academic professions, often in very high positions," was almost inevitable given the autists' deep passions and keen intellects:

> Able autistic individuals can rise to eminent positions and perform with such outstanding success that one may even conclude that only such people are capable of certain achievements. It is as if they had compensatory abilities to counter-balance their deficiencies. Their unswerving determination and penetrating intellectual powers, part of their spontaneous and original mental activity, their narrowness and single-mindedness, as manifested in their special interests, can be immensely valuable and can lead to outstanding achievements in their chosen areas. We can see in the autistic person, far more clearly than with any normal child, a predestination for a particular profession from earliest youth. A particular line of work often grows naturally out of their special abilities.

From this description of autism, finally available to English-speaking audiences more than forty-five years after it was first published, the gap between prodigy and autist seems quite small. With minimal tinkering, the same passage could have been written about child prodigies.

But some researchers questioned whether individuals like those described by Asperger were actually autistic. When Asperger's disorder first appeared in the *DSM* in 1994, it was listed as a diagnosis *separate from autism*. It wasn't until the *DSM-5* was published in 2013 that Asperger's disorder was formally enveloped into autism spectrum disorder.

Encompassing Asperger's disorder within the folds of autism was part of a larger trend toward loosening the definition of autism. Even before the *DSM-5* was issued, autism diagnostic criteria had been evolving in ways that led to more autism diagnoses and to people with a broader range of abilities being included on the autism spectrum.

New evidence also made autism and intellectual disability seem less closely intertwined. A 2006 review study challenged the evidence on which claims of extensive overlap between the two conditions had been based. Another study found that while a significant percentage of autists were intellectually disabled, a significant percentage also had average IQs, and some even had above-average IQs (a finding that might have been due in part to the broadening criteria for autism).

Another line of research suggested that merely swapping out an intelligence test that required oral instructions and responses (the Wechsler Intelligence Scale for Children) for a differently structured, nonverbal test (the Raven's Progressive Matrices) significantly increased autists' scores. In one study, switching to the nonverbal test catapulted the autists' average score thirty percentage points and removed the scores of all but 5 percent of the autistic children from the "low functioning" range.

The cognitive theories of autism, too, evolved to more fully incorporate autistic strengths. The executive function theory (based on the idea that autism stems from deficits in planning, goal-setting, and related abilities) was essentially dismissed as a primary explanation for autism.

The weak central coherence theory (based on the idea that autists have an imbalance in the way they integrate information) was revised to emphasize autists' superior local processing rather than their failure to see the big picture.

Baron-Cohen recast mind blindness as the empathizing-systemizing theory of autism. The new theory emphasized that autists had intact or strong systemizing, the drive to find the rules that govern systems such as language syntax, train timetables, and tidal wave patterns. Though an empathy deficit remained at its core, even this component of the theory didn't divide prodigies and autists as much as it might seem. Baron-Cohen and other researchers believe that empathy actually has two components—cognitive empathy (the ability to recognize the feelings of another) and emotional or affective empathy (having an appropriate emotional response to others' feelings). These researchers think that autists have a deficit in cognitive empathy but have intact or an overabundance of affective empathy. Baron-Cohen has further explained that even though autists lack cognitive empathy, they are often "supermoral." The drive to systemize, Baron-Cohen says, leads autists to develop highly sophisticated moral codes and a strong sense of justice.

These new and revised theories portray autism as a condition that encompasses strengths as well as weaknesses. The revised weak central coherence theory emphasizes autistic attention to detail, a trait shared with the prodigies. The empathizing-systemizing theory of autism accounts for autists' strengths in rule-based subjects, like math and chess, and specifies that autists have excellent affective empathy—just like the prodigies. The intense world theory portrays autists as highly empathetic and perceptive individuals with enhanced memory capabilities. These strength-recognizing theories, built from the rubble of the early, deficit-focused theories, offer a starkly different take on autism. From this perspective, the connection between autism and prodigy isn't just conceivable; it's almost inevitable.

Chapter 8

Another Path to Prodigy

Sometimes the genetic, autism-linked explanation for prodigy comes up short. The family connection between autism and prodigy is strong, but it's not perfect. Some—roughly half—of the prodigies come from families without autistic relatives.

These prodigies still have autistic traits. They demonstrate the same heightened attention to detail and penchant for obsessive interests as the other prodigies.

How could these traits stem from a family link with autism if there's no autism in the family?

There are a few potential explanations. Perhaps there's autism in far-flung parts of the family tree. Perhaps it's the unique combination of the prodigies' parents' genes that introduced some autism-linked traits into the family for the first time. Perhaps these prodigies didn't inherit the relevant genes at all but have de novo genetic mutations— mutations present in the individual but not in either parent—that contribute to their incredible memories and focus.

But there's another possibility as well. Perhaps the pathway to prodigiousness is paved not just by the children's genes but also by their environments—the events or substances to which they are exposed prenatally or even after birth.

This seems to be the case for autism. Though it's highly heritable, genes don't always tell the whole story. There are some known environmental risk factors for autism (environmental in the sense that they aren't directly tied to genes—not in the long-discarded "refrigerator mother" sense). Children exposed in utero to valproate, an anti-epilepsy medication, or to thalidomide, a medication once tragically prescribed to treat morning sickness (and now used to treat skin conditions and cancer) and known to cause severe birth defects, have increased rates of autism. Similarly, some studies have demonstrated an increased risk of autism for children with congenital rubella (a condition that can develop when a pregnant woman contracts rubella).

Exposure to such environmental risk factors doesn't always actually result in autism. What accounts for the variance in outcome? Some scientists have proposed that one factor is the individual's genes; some people's genes may leave them more susceptible than others to such environmental exposure. Their genes and the environment interact in a way that may result in autism—though the same would not necessarily be the case for others with the same environmental exposure and different genetic profiles.

Perhaps, just as with autism, there are environmental "risk" factors for prodigy—events or exposures that increase the likelihood that a genetically predisposed child will demonstrate the unchecked drive, incredible memory, and heightened attention to detail that characterize prodigious behavior.

It's a possibility that the Tiessens, a Canadian family of four with first one and then two prodigious sons, have experienced firsthand.

In 2012, *The Huffington Post* featured Josh Tiessen as one of "ten art prodigies you should know." The accompanying video shows an interview with the artist, a soft-spoken seventeen-year-old with carefully styled, slightly spiked brown hair. He appears very thin, almost gangly, and he speaks with gentle reverence about celebrating God's creations through his art.

The video of that artwork reveals stunningly detailed pictures of animals and architecture: in *Snow White,* a portrait of a tiger Josh created at thirteen, he painted tiny, distinct hairs on the animal's face and body; in *Behold the Door,* a zoomed-in view of a battered doorway Josh painted at fifteen, he carefully portrayed chipping paint, knots in the wood, and tiny nails.

Joanne met Josh in the winter of 2013. Josh had graduated from high school but still lived with his parents in Ontario. Joanne drove through a snowstorm to the family's charming Tudor-style home. When she arrived, Josh's parents, Julie and Doug, provided a detailed family history. No one in Doug's family was autistic. Julie was adopted; she had some information about her biological parents' families, but there were also blank spots in the family tree. As far as Julie and Doug knew, though, they didn't have any autistic relatives.

Julie and Doug brought Joanne to Josh's studio—a bright space at the back of the house with a slanted ceiling and shiny wood floors—and showed her some of Josh's work. It was as extraordinary as it appeared online.

Joanne knew that the Tiessens' younger son, Zac, had a talent for music, but it wasn't until Doug and Julie walked Joanne through his history that Joanne realized he, too, seemed prodigious. He had the same lightning-quick development of a skill that had served as Joanne's hallmark for these distinctive children.

But, unlike the other prodigies, Zac didn't show any particular passion for music early on. As a child, he hated music class and refused to take up an instrument. It wasn't until a thirteen-year-old Zac bashed his head against a church floor that he wanted anything—and then everything—to do with music. Up to that point, he was just the kid brother of a child prodigy.

⟶

Josh Tiessen was born in Russia, where his parents were then serving as missionaries. It had been a harrowing pregnancy and birth that slung his parents through a labyrinth of Russian hospitals and medical procedures, but Josh emerged seemingly unscathed.

Zac eased into the world in a Moscow hospital thirteen months later, the product of an uncomplicated pregnancy and birth.

Josh learned to hold a crayon around the time Zac was born. He immediately wanted to draw, but he ignored the stacks of coloring books his grandparents mailed to Russia; he only wanted blank paper. His nanny, Lena Zhyk, fussed over Josh's artwork. Look what he did today, she would say while holding up a sheet of paper. Julie and Doug rolled their eyes. It's just scribbling, they thought to themselves. There's no picture there.

When Josh was around three, Lena taught him perspective. She held up stuffed animals for him to draw and talked to him about shading. Josh would sit at the small table in the playroom drawing for long periods of time, sticking his tongue out of the corner of his mouth as he worked. Lena often corrected him; if Josh got the perspective wrong, she would rub out what he had done and draw or paint over it.

Zac occasionally scribbled in the cast-off coloring books for a few minutes, but no project held him for long.

Increasingly impressive artwork began emerging from the playroom, but Julie was convinced Lena still had a hand in its creation. By the time Josh was five, Lena swore she was no longer touching his projects. Julie and Doug were skeptical that they had a great artist on their hands, but they encouraged Josh to sign his pictures and jokingly referred to them as "Joshy originals." Colleagues who visited the Tiessen home never believed the pictures were truly Josh's work: the art was far too advanced for five-year-old hands.

When Josh was six, the Tiessens moved back to Canada. For several years, Josh's interest in art fizzled, at least as far as Julie and Doug could tell. School occupied much of his day. Sometimes, though, when Julie thought Josh was playing, she would find him up in his room drawing sports logos or sneakers at his desk; when he was watching TV, he would whip out paper and begin sketching. But the long afternoons he had spent on art as a toddler seemed to have been left behind in Russia.

After Josh finished third grade and Zac second, Julie began home-schooling the boys. Josh took to it immediately. He loved the quiet atmosphere and the wide-open afternoons no longer stuffed full of classes and activities.

Zac was a dervish of a student. "We had lots of blood, sweat, and tears that first year trying to get Zac to concentrate," Julie recalled. "I cried a lot of days."

One day, frustrated with the boys, Julie gave Josh and Zac paper and pencils and told them to go outside and draw. Zac spent fifteen or twenty minutes sketching something that vaguely resembled a fountain and then abandoned the project. He spent the rest of the afternoon underfoot in the house.

Josh perched himself on the family's lawn and set about drawing the long, rambling home they lived in at the time. He zeroed in on every brick, every roof shingle, the design on the doors. When he ran out of space, he came back into the house for more paper, eventually taping several pieces together. Julie and Doug had to coax him in for dinner.

When Julie found a book on perspective at a library shared by a group of homeschooling parents, Josh read it, studied it, and incorporated what he learned into his drawings. Zac quickly got bored with the project and moved on, creating havoc in the home classroom.

Julie enrolled both boys in a church arts club to complement her homeschooling lessons. One of the club advisers, Valerie Jones, a British expat in her mid-sixties who crafted animal portraits as a serious hobby, noticed Josh immediately. She watched as the nine-year-old used clean strokes and hard lines to embed his name within a red-and-blue geometric design on a name tag; he was completely engrossed in his work.

Over the next few weeks, she kept an eye on Josh as the group worked on shading and perspective. The other children all produced what Valerie thought of as "kids' art," but not Josh. His drawings had incredible precision.

His approach to the projects was different, too. Other kids grew distracted; they couldn't focus on any one task for too long. The room, filled with ten or so kids, was busy; it sometimes got loud. But Josh shut everything else out. He focused intently on executing his drawings.

Valerie was convinced that this was a talent that needed to be nurtured. She sought out Josh's parents and raved about their son's abilities. She later invited Josh and Zac to her studio for lessons, and the boys began spending Wednesday afternoons in Val's basement studio, working at a table adjacent to her laundry room. Zac continued for six months or so before giving it up. Josh kept at it. He was quiet, respectful, and surprisingly mature—almost like a miniature adult. He soaked in Val's instructions; he was never distracted.

Soon, Val was ushering Julie down to the laundry room to see Josh's latest projects. Julie was shocked at what she saw, particularly when Val showed her Josh's chalk pastel depiction of a lion inspired by Aslan, a Christlike figure from the *Narnia* series. The lion's green eyes appear liquid, his mane wild; he possesses a stirring dignity. In a moment reminiscent of Josh's toddler days with Lena, Val insisted that Josh had done it entirely on his own.

After a few months, Val called Julie to ask if she could arrange an art exhibition for Josh. Julie laughed. Josh was only ten! But Val insisted that the world needed to see his work. Julie relented, and Josh had his first exhibition at eleven and his first sale to a stranger: a nurse purchased one of his photographs, a shot of a child taken during a mission trip to Honduras.

Other exhibitions—and sales—trickled in. Josh displayed his work with other artists at a local business, a church, and a gallery. At fourteen, he had his first solo gallery exhibition when his work was featured on the Community Wall at the Art Gallery of Burlington.

Zac attended every event. He was always a spectator, never a participant. He helped carry canvases and, in exchange for a small commission, sold artist note cards. He never showed anything of his own.

Josh got his big break when he grabbed the attention of Robert Bateman, a prominent Canadian painter. He was Josh's professional idol—a man who, much like Josh, specialized in detailed, realistic portrayals of nature and animals. Josh had written to him a few months before his Burlington exhibition, attaching images of his paintings. Just a day after Josh took down the last of the works he had hung at the Burlington gallery, he got a response. Robert praised Josh's work and invited him to a Master Artist Seminar.

The seminar was on Cortes Island, off the coast of British Columbia, more than twenty-five hundred miles from the Tiessens' home. Money was scarce. Doug had been diagnosed with chronic Lyme disease two years before, a condition his doctors believed he had contracted in Russia, and his health was failing. Julie's health was declining as well; she would be diagnosed with chronic Lyme disease the following year. Long-term disability payments were the family's only source of income.

Josh's aunt and uncle volunteered their frequent-flier miles. Josh and his parents scrambled to find money to pay for the rest of the trip. Through a scholarship from the center hosting the seminar, an unexpected contribution from a family friend, and money donated by the Kiwanis Club in exchange for a painting, the Tiessens pieced together the funds to send Josh and Julie to British Columbia, and Josh attended the seminar.

Infused with the golden endorsement of Robert Bateman, Josh's career took off, and the media attention increased. His price tags shot up, too. Bateman advised Josh that work of his caliber could sell for much more than he was charging, so the fifteen-year-old upped the asking prices for his originals into the thousands.

Zac had constant exposure to the Josh Tiessen art frenzy, and, as Julie puts it, for years his "nose was kind of out of joint" about the whole thing.

Josh began pouring more and more time into his art. For him, that time was bliss. He prayed before he began and played music or listened to lectures on faith and art while he painted. Everything else was a distraction.

The common ground between Josh and Zac eroded: Josh frequently opted out of their hour-long daily allotment of TV; he lost interest in gaming. The brothers had previously played basketball together, but after they enrolled in a Catholic high school, neither boy made the school team.

At fifteen, Josh set up the Josh Tiessen Studio Gallery in the sunroom. It was a massive upgrade in work conditions. Josh had previously done a stint in the garage. As winter approached, he had moved a series of heaters out to his studio, but none beat back the frigid air. His next move had been to the basement laundry room. It was warmer, but paint splattered the Tiessens' washer and dryer. As collectors began visiting, Josh felt awkward bringing them down to his makeshift work space.

Just as Josh was turning sixteen, he graduated from high school. Over the next few years, the achievements rolled in. Josh pocketed a series of honors at local art festivals and competitions for teens. He nabbed the second most votes in So You Want to Be an Artist, a national Canadian contest for teens, for his close-up depiction of an intricate doorknob on a weathered door; the painting was then displayed at the National Gallery of Canada in a monthlong exhibition. The famed Canadian conductor Boris Brott invited Josh to create a piece to accompany one of his symphonies, an honor Josh performed twice. He was invited to join Artists for Conservation and the International Guild of Realism. He was one of sixty thousand Canadians to receive a Queen Elizabeth II Diamond Jubilee Medal, an award given to individuals with notable achievements or contributions to their communities.

That contribution included not just his art but also a striking benevolent streak that Josh shared with the other prodigies. He gave a portion of his earnings to charity, frequently donated artwork for fund-raisers, and initiated an annual artists' event with a charitable purpose. He launched his own foundation, Arts for a Change, to coordinate his philanthropic work.

When others inquired about the source of Josh's talent, the source of his passion, Julie and Doug, two missionaries who had never considered themselves particularly artistic, always gave the same answer. *He just came this way.*

⌒

Zac's childhood looked a bit different.

He was active from birth. The nurses at the hospital nicknamed him Houdini for his ability to escape from even their tightest swaddles. While a toddler-age Josh pored over art projects, Zac exasperated Lena by skipping from toy to toy, pushing around cars and trucks. He experimented with Lego bricks while his older brother mastered drawing with perspective. No toy was safe in his hands; playthings that had survived multiple other children broke within weeks.

He was always in motion, always on the verge of catastrophe. He loved to climb, and Lena was constantly rushing to pull him off things—ladders, chairs, tables—trying to grab him before he toppled. At a friend's apartment, Julie once found Zac perched on the ledge of an open ninth-floor window. Another time, he got his head stuck between the metal bars of a porch railing in Russia and screamed for twenty minutes before Lena, Julie, and Doug wriggled him out. When Julie and the boys joined some of their Russian friends for a picnic near the Black Sea, Zac kept trying to run off a cliff. Julie joked that if she could just keep Zac alive until he turned five, he would be fine.

He was a social creature and craved the company of other children in a way that was foreign to his older brother. Julie ran a one-room schoolhouse for her boys and a couple of other families during their last year in Russia. It was a torment to Josh, who had nightmares about other children destroying his toys, but Zac exalted in the mayhem.

Zac's parents always suspected he had a talent for music. He hummed when he was playing, eating, and falling asleep; he hummed in the car and in the bathroom. When Julie or Doug asked about the tune, Zac would cite the background music he had heard earlier in a restaurant or store or on a commercial—music Julie and Doug hadn't even noticed. He had an excellent sense of pitch and a good singing voice. Lena and Josh warbled their way through the Russian songs Lena taught the boys; Zac was the only one to hit the notes. Julie often wrangled Zac into sitting next to Josh and singing into his ears to help him stay on key.

Despite his seemingly natural gift, Julie and Doug could never pin Zac down long enough for him to develop it. Music classes at school repulsed him. As he would later recall, nothing sounded right. The other kids' singing was erratic; the piano was out of tune. When Julie began homeschooling the boys, she tried to integrate music into the curriculum, but Zac hated it. He fidgeted when sight-reading music or singing from hymnals. He couldn't be bothered to learn about the history of music or classical composers.

Zac's lack of interest frustrated Julie and Doug. Music, they felt, was a place Zac could excel—a route for him to develop a talent of his own. Julie and Doug took Zac to a music store before Christmas and tried to entice him into asking for an instrument. No dice. Someone gave him a toy piano, and his parents bought him a toy electric guitar. Neither took.

Zac flitted from activity to activity. If any project or pursuit lasted long, he lost interest. One year, Josh and Zac set out to build a large fort in a ravine, but Zac abandoned the project. Doug occasionally built model cars with Zac, who would help out at first but then leave Doug to finish the models alone. Zac and Josh joined a Bible quiz group, and Josh diligently memorized verses while Zac struggled to put in five to ten minutes a day. When Josh began devoting hundreds of hours to more complex pieces of art as a teenager, Zac watched TV and played video games. His parents suspected he had at least a borderline case of ADHD.

If there was fun to be had, though, Zac was in the thick of it. He was at the heart of every group; he emerged from every activity with a new best friend. He passed long afternoons gaming or playing Ping-Pong with friends, and he issued a constant stream of invitations to the Tiessen home, always looking for someone to hang out with.

Until the day Zac slammed his head against a church floor. After that, he was different.

↝

It happened when Zac was thirteen.

Zac's church youth group meeting ended, and most of the kids headed out to the foyer to wait for their rides. Zac and some of the others tracked down a few empty appliance boxes they had used as part of a game during the meeting. He and his buddies flattened one out. They took turns lifting it and jumping over it, raising the make-shift hurdle a bit higher each time.

On Zac's turn, he dove headfirst over the box. He cracked his head against the thinly carpeted concrete floor. For a minute, maybe a minute and a half, Zac blacked out.

When he regained consciousness, the youth group leaders helped him to the side of the room and leaned him against the wall. Some-one brought him a glass of water and placed it next to him on the floor. Zac, still dazed, reached for it. He knocked it over. Someone ran outside to get Doug, who was waiting in the parking lot.

Doug had been through this before with Zac; the kid couldn't play Ping-Pong without making it look like an extreme sport. His stunts often landed him in the hospital, and he'd already had multiple head injuries and concussions (his parents joked that he had a "club card" for the emergency room). Doug trotted out his familiar list of questions. Zac didn't know his name. He didn't know his birth date or address. He insisted that five plus five was twelve. He was worse than Doug had ever seen him. Doug and Josh helped Zac out to the car. Zac felt overwhelmed with fatigue; all he wanted to do was sleep, but Doug made him stay awake for the drive back home.

Julie knew the drill. She checked the head-injury guidelines that the family had posted inside the medicine cabinet after Zac's previous concussion, a tobogganing injury that led to an overnight hospital stay. Zac's pupils were still dilated, and he was still nauseated. But he was a bit more coherent than he had been right after the accident. Julie and Doug decided to skip the trip to the emergency room. They woke Zac every couple of hours and observed him through the night. He woke easily enough, and Julie and Doug assumed that all was well.

The next morning, Zac seemed more low-key than usual. Julie and Doug wrote it off as the aftermath of interrupted sleep. They let him stay home from school for a couple of days.

When he went back, he was reclusive. Before the accident, he had always been friendly with his classmates. Afterward, he just wanted to be alone. Over the next couple of weeks, Julie noticed that Zac was quieter than usual; he went out less. At his youth group meetings, he wasn't as wound up, not so much the life of the party. Maybe this injury finally got through to him, Julie thought. Maybe Zac realized how serious a concussion could be.

A few weeks after Zac's accident, the family had Valerie, Josh's painting mentor, over for dinner. Valerie knew that Zac hadn't made the high school basketball team—the one activity he had stuck with for more than a short period—and she imagined it had to be difficult watching Josh's achievements from the sidelines. Hoping to raise his spirits, she had picked up a guitar for him at a flea market. "Maybe it'll be something for him to tinker on," she had said to her husband.

When she gave it to Zac, he was intrigued. He took the guitar out into the living room and began strumming. He worked his way across the notes slowly, as if he were exploring the instrument. He kept at it all evening. "We were all laughing," Julie said. "We had never seen him do anything this long."

After dinner, Zac ran right back to the guitar. He harassed Julie, who had played as a teenager, for her old guitar books. When Julie couldn't find them, Zac searched for instructions online. After Val left,

Julie showed him some chords. Zac soaked up every movement. Hours passed; eventually, all the other lights went off in the house, and Julie told Zac to go to bed. The next day, Zac went straight back to the guitar. Julie was pleasantly perplexed by Zac's sudden diligence.

For a couple of months, Zac practiced for an hour or two a day.

He pestered his parents to let him use the money he made doing yard work to buy an electric guitar. Doug and Julie were skeptical. They had been through a laundry list of activities with Zac. They didn't want him to buy an expensive instrument that he would discard after a few weeks. But when Zac's interest didn't waver after three months, they relented. Doug helped Zac find an electric guitar and amp on a used-goods Web site. The next week, Zac bought the levels one and two FastTrack guitar instruction books at a used-books sale.

Julie helped him work through most of the level-one book; she still knew a dozen or so chords. But Zac quickly surpassed Julie's abilities. He raced through the second book on his own. "I felt really connected with the music and the instrument," he recalled. "It just kind of naturally happened."

He started practicing at least three or four hours a day.

As his technical skills improved, he sought out tough, complicated songs. He grew infatuated with heavy metal bands ("screamo" groups, as Julie calls them). But Zac, who by this point wore dark-framed glasses and kept his dark hair long, was drawn to the fast, demanding guitar solos.

A drive for perfection, a quality the Tiessens had never before thought of as very Zac-like, kicked in. Zac set his sights on recording a complicated, lightning-fast classical piece. He cranked up the tempo for his performance, then posted the video on YouTube. Zac quickly grew dissatisfied with it. His performance wasn't clean enough. The sound quality of his equipment wasn't good enough. He took the video down.

Zac zoomed in on music until there was little else left in his field of vision. Julie and Doug insisted that he go to church and attend youth group, so he did, but other activities vanished. Gaming fell by the wayside; skateboarding disappeared from his routine. It became a fight to get him out of his room for dinner. Julie and Doug would find him at 3:00 a.m., 4:00 a.m. still awake, still playing guitar.

He stopped inviting friends over, and he stopped seeking out social activities. He had other kids over to jam a couple of times, but it was a bust. The other kids weren't serious enough; they just wanted to fool around. He joined a youth band at church, but that was no better. The other kids skipped practice or missed their notes. He switched to the adult worship band that performed during church services, but his role was restricted to quietly strumming in the background. When he grew frustrated with his inability to perform the fast—sometimes frantically paced—pieces he loved, he persuaded his family to switch to a more arts-focused congregation.

He started practicing four to six hours a day.

About a year after Zac first picked up a guitar, he turned to composing. He borrowed stacks of guitar and music theory books from the library. When Julie peeked into his room, she found Zac, a boy who had never before shown any interest in books, poring over the thick volumes. In disbelief, she watched him turn the pages. At fourteen, he began studying music theory through ABRSM, a UK-based organization that provides music-learning resources and exams. He worked through the first six grade levels in a year.

When he tried his own hand at composing, he started with metal and classical music, then gravitated toward a more progressive style, a type of music he describes as "anywhere from metal to jazz, and classical to flamenco, all combined." He was obsessed with creating music. He stayed up late. He skipped meals. The family took a road trip through New Brunswick and Nova Scotia, and Zac barely left the car; he played and composed the entire trip.

Ideas came to him all the time. If he was out of the house, he stopped what he was doing to scribble notes to himself. As soon as he got home, he went directly to his room to work out his new idea. He woke in the middle of the night, seized by the need to change one or two notes in a composition.

At the Christian bookstore where he worked, Zac was still polite. But at home, he grew moody. He suffered from headaches and confusion. If his parents tried to enforce breaks from music, he could even become combative. It might have been a product of all his concussions (his family has lost track of the total count, but they estimate that he's had at least ten). It might have been too much working without a break. It might have been a symptom of the Lyme disease that the family had recently learned that Josh and Zac, too, had contracted. During the worst bouts of behavior, Doug and Julie insisted that Zac hand his laptop over at midnight. They forced him to rest. When his mood improved, they would ease some of the restrictions, and Zac would again immerse himself in music.

By the time Joanne met Zac, it had been just over three years since his youth group concussion and since Val had given him his first guitar. In the interim, he had taught himself to play the six-string electric guitar, the balalaika, and the baritone ukulele. He had recently gotten an eight-string electric guitar, and it was becoming his primary instrument. He had enrolled in the Berklee College of Music distance-education program and snagged a scholarship from the Kiwanis Club to attend a musicians' workshop taught in part by Tosin Abasi and Javier Reyes, master eight-string guitarists. He had crafted sixteen full-length songs. He composed for the instruments he played and for those he had never learned—drums, French horn, cello, and alto sax.

The boy whose behavior had seemed touched with ADHD for the first thirteen years of his life was engrossed in music six, eight, or even twelve hours a day. His parents had no idea what had happened.

⟜

The link between Zac's concussion and his altered behavior may seem clear in hindsight. But the Tiessens had been through other concussions with Zac, and those hadn't had any long-term impact. At the time, Julie and Doug didn't see Zac's head injury as a potential cause of the changes in their son. It wasn't until Joanne connected the dots between Zac's concussion and the dramatic unleashing of his music that the Tiessens realized that Zac's injury might have triggered his altered behavior.

This realization prompted soul-searching about the source of *Josh's* prodigiousness. Could Josh, too, have suffered an injury? Retracing Josh's childhood left the Tiessens empty-handed. There were no notable injuries to report. But his days in utero were a different story.

Julie had endured bleeding and extensive preterm labor with Josh, including excruciating contractions. Medical care in Krasnodar, the city in southwest Russia where she and Doug were stationed on their mission, was limited. Several times she flew to a hospital in Moscow for care. She was pumped full of unfamiliar drugs—drugs to stop the contractions, drugs to stop her bleeding, drugs to help her sleep. For months, a miscarriage seemed imminent.

Then there was Julie's terrible fall ten days before Josh was born. During Julie's fourth prenatal trip to Moscow, the doctors insisted that she stay in town until she gave birth. The Tiessens holed up at their mission's Moscow guest apartment and settled in to wait for their baby's arrival.

From the beginning, there had been unusual activity at the apartment. Every couple of days, someone called. It was usually a man and always a Russian speaker. The person would refuse to give his or her name. He always demanded the address of the apartment.

Once, while Julie was home alone, a man repeatedly rang the doorbell. When she refused to let him in, he finally left. Half an hour later, the phone rang. Julie answered and heard heavy breathing before the line went dead. A few minutes later, the phone rang again. This time, a man said that he was a friend from Krasnodar who had brought her presents and souvenirs and asked why she hadn't let him inside her apartment. He rattled off Julie's address. Julie told him not to call back, and she hung up the phone.

Two days later, Julie was at home with Doug and his mother, who had come from Canada to help the couple prepare for the baby. Doug and his mother had been planning to run errands, but at the last minute both stayed home to wait for a courier delivery for another missionary. When the doorbell rang, Doug answered.

A black-gloved hand shot through the opening and grabbed Doug by the neck. The intruder tried to push his way in. His head was wrapped in white cloth, completely covered except for small slits cut out around his eyes. Doug shoved back against the door. Julie sprang up from the kitchen bench to help him, but she tripped and fell on the rug that ran the length of the hallway, landing on her stomach.

Doug slammed the door shut. The intruder screamed. Julie pulled herself up and ran to the window. She saw a large man clutching a white bundle rush outside to a car. It was too far away to make out the license plate number.

A detective came to talk to the Tiessens. The Tiessens knew enough Russian at that point to chitchat with a neighbor, but police lingo was beyond their reach. They caught the word "ring" and inferred that the detective was describing some sort of crime ring.

The mission paid to have a stronger door installed in the apartment complex. There were more disturbing calls, and Julie panicked every time the doorbell rang. But they never saw the attacker again, and Julie was relieved that at least her fall hadn't induced labor.

When Josh finally emerged a little over a week later, he was a bit lethargic, a bit jaundiced, but otherwise seemed no worse for wear. But could he have been *different* for the wear? Could all of that in utero trauma have altered something in his brain?

↤

The mystery of Josh and Zac Tiessen's prodigious abilities cries out for a brain scan.

Less than a year before Joanne met the Tiessens, David Feldman closed a talk he gave at New York University with a plea for someone to take up the baton and run a brain imaging study on child prodigies; there had not yet been any done. "Please fix that," he requested.

There were murmurs of interest from the crowd, but Feldman heard a lot about the obstacles. The machines were expensive to run. It was hard to get funding to study prodigies: there was nothing wrong with them, they didn't need help.

Feldman made the argument, based on Joanne's work, that studying prodigies might lead to advances in autism research, but nothing ever came of it. Feldman and a collaborator applied for a grant to do the work themselves, but their application was turned down. The inner workings of the prodigy brain remain a black box.

Once again, there is something to be learned from savants, those individuals whose skills—sometimes prodigious in nature—coexist with disability. In savants, the appearance of spectacular skills following an environmental trigger, like an illness or an injury, is well documented. These individuals are known as acquired savants, and Darold Treffert, the Wisconsin savant expert, estimates that they make up more than 10 percent of the savant population.

Their stories feel ripped from the headlines. A normal life. An injury or illness. The spontaneous eruption of a new ability and passion.

Jason Padgett was jumped after walking out of a karaoke bar and took multiple blows to the head. Prior to the attack, Jason had never been interested in math. He was a college dropout whose math studies had gone no further than pre-algebra. Afterward, he rattled off prime numbers in his sleep and couldn't stop counting. He envisioned pi and mathematical equations as fractals, and he drew the highly complex images they evoked for him.

Alonzo Clemons fell and hit his head while jumping from the toilet to the bathtub as a toddler, an injury that left him unable to read or write. Soon thereafter, he developed an insatiable need to sculpt. He used whatever he could find around the house—including soap and shortening—as his modeling clay, and he crafted incredibly detailed, lifelike animals. He sculpted through an interview with Morley Safer for *60 Minutes* and while waiting in the greenroom to appear on *The Morning Show* with Regis Philbin and Kathie Lee. Once, while Alonzo was living in a state training school in Colorado, staff members took away his clay, planning to return it to him as a reward for conquering activities that were much more difficult for him, such as speech, combing his hair, and tying his shoes. Soon after, school staff members found tar streaked across Alonzo's bedding. They inspected his room and discovered a host of small, sticky black animals beneath his bed. Stripped of his clay, Alonzo had scraped tar from the pavement and the edges of windows to make these figurines. He had to sculpt, and he had to do it all the time.

Tommy McHugh never showed any interest in the arts until, at fifty-one, he had brain aneurysms that led to a stroke. When he awoke, he spoke in rhyme and began writing poetry. He generated hundreds of sketches and drawings and began painting anywhere he could find a surface: on canvases and on the walls, floor, and ceiling of his house. He often painted faces, images he once described "as his personality crying for help to save him from his obsession."

Acquired savants have been the subject of much more research than prodigies. By examining their cases, scientists have begun to piece together just how an injury—an event usually tied to a *decrease* in functionality—could lead to *improved* abilities.

At least as early as 1980, researchers began theorizing that perhaps these sudden outbursts of talent have something to do with injury to the brain's left hemisphere. One case prompting such speculation was that of Mr. Z., a man who had been shot in the left side of his head as a child during a robbery of his home in rural Mexico. The injury initially left him mute, deaf, and partially paralyzed. A couple of years later, he regained his hearing and the ability to walk, though his speech remained impaired; he also displayed impressive mechanical skills, including dismantling and reassembling bicycles, designing a punching bag that mimicked the actions of a human opponent, and reproducing pictures with impressive accuracy.

Others had already observed that many savant skills were rooted in the nondominant hemisphere of the brain—typically the right hemisphere. A psychologist examining Mr. Z.'s case suggested that perhaps his left-hemisphere injury had resulted in right-hemisphere "overcompensation"—an overdevelopment, of sorts, that led to the emergence of his spectacular skills.

It was a line of thought that gained traction when, beginning in the mid-1990s, the neurologist Bruce Miller and his colleagues documented surprising skills in dementia patients. One such individual, a former businessman, lost much of his memory and verbal skills; he changed clothes in public, scoured the sidewalk for coins, and shoplifted. He had no previous interest in art—he had never even visited a museum—but he quit his job to take up painting. His work evolved from colorful ellipses to increasingly detailed portrayals of animals. He won awards at local art shows.

This former businessman had a specific type of dementia—frontotemporal—that attacks certain parts of the brain while leaving other areas relatively intact. Miller and his colleagues eventually calculated that 17 percent of their frontotemporal dementia patients demonstrated new abilities or preserved preexisting artistic or visual abilities despite their worsening conditions.

It was a shocking phenomenon. Miller noted to a reporter for the *Washington Post* that it had never occurred to him "that somehow a disease could release an unknown talent."

These findings, moreover, dovetailed with the left-brain injury, right-brain compensation theory. Miller and his colleagues discovered that among their talented frontotemporal dementia patients, the disease had generally attacked their temporal lobe but spared their frontal lobe, and that most such patients showed greater deterioration of their *left* hemisphere than their *right*.

Researchers put forward a flurry of theories to explain how left-hemisphere brain damage might result in savant-like abilities. Some proposed that damage to the left hemisphere might lead to increased development of the right. Others thought that left-hemisphere damage merely *unleashed* latent right-hemisphere abilities previously held in check by the dominant left hemisphere.

Whatever the precise mechanism, the left-brain injury, right-brain compensation theory suggested that savants weren't as dissimilar from other people—their talents might not be quite so inimitable—as it seemed. This raised an intriguing question: If scientists could simulate the left-hemisphere damage found in many of the talented dementia patients and some savants, could they expose the inner savant within us all?

Allan W. Snyder, the founder of the University of Sydney's Centre for the Mind, gave it a try. Snyder and his colleagues ran a series of experiments in which participants—typical, non-savant individuals—donned a "creativity cap" of sorts. The experimenters attached electrodes to the participants' brains and zapped them with a weak electrical current targeted at inhibiting part of the left temporal lobe, an attempt to mimic the left-hemisphere injuries of the acquired savants.

With part of the left hemisphere inhibited, at least some participants in each study showed an improved ability to proofread and differentiate true memories from false memories. Some showed stylistic changes in their drawing; some demonstrated improved numerosity, the ability to gauge the number of items in a group (the classic example of this was documented by Oliver Sacks when a pack of matches spilled and twin savants immediately cried out "111"—the number of matches on the ground). In a slightly different experimental design in which the researchers not only inhibited part of the left hemisphere but also stimulated part of the right, more than 40 percent of participants were able to solve a difficult puzzle that had previously stumped them.

But not everyone who sported the creativity cap showed improvement. Many people still couldn't solve the difficult puzzle, more than a third showed no reduction in false memories, a majority showed no change in drawing style, and a couple of people showed no improvement in numerosity. Similarly, not all patients with frontotemporal dementia develop or preserve special skills. Nor, certainly, does everyone who suffers a left-hemisphere brain injury or illness emerge a savant.

There are at least a few ways to explain these inconsistencies. Perhaps the focal point of the electric shock Snyder and his colleagues administered landed off target and failed to inhibit the relevant part of the left temporal lobe. Perhaps not every head injury exposes savant skills because such transformation requires an extremely precise—and rare—type of injury. Perhaps there is another factor that prevents some individuals with frontotemporal dementia from exploding with artistic or musical interest.

But it seems at least as plausible that the premise is wrong; maybe not *everyone* has an inner savant. Or maybe that inner savant is much more difficult to expose—for potentially interesting reasons—in some people than in others. Just as not every child with prenatal exposure to valproate or thalidomide is born with autism, it seems that not everyone who incurs a head injury (or dons the creativity cap) will develop prodigious skills. If true, that would suggest that some people are more susceptible to savantism than others. The interesting question then is why.

There's a very limited number of studies in which scientists have attempted to induce savant skills, and those studies that have been done are quite small, making it difficult to identify any characteristics that might distinguish those who demonstrate such skills from those who do not. But a 2004 study in which a trio of researchers in Australia applied the brain-zapping protocol and then looked for savant skills to emerge across a variety of areas identified one potentially important factor—gender. The male participants, the authors found, generally performed better under stimulation than the female participants.

This finding hasn't been replicated, and in one other small study, gender wasn't predictive of the impact of the creativity cap. But while it's certainly possible that the 2004 finding was a fluke, it seems worth contemplating whether gender impacts the ease with which savant skills can be induced. After all, the ratio of male to female savants is similarly skewed, possibly as lopsided as six to one. There's a similar gender breakdown in Miller's skilled dementia patients, among autists, and among prodigies. It seems that men may be more vulnerable than women to each of these conditions.

There's at least one theory as to why this might be the case. Studies have found that elevated levels of testosterone and other hormones in utero are associated with an increased risk of autism, and that boys naturally have more such exposure. Could this type of exposure also leave the brain primed for savantism and prodigiousness?

⟜

There was something odd about Jason Padgett's brain scan. When Padgett was presented with mathematical formulas during an fMRI, his results showed left-hemisphere activation—the opposite of what the left-brain damage, right-brain compensation theory would predict. As the authors of that study noted, his results were "perhaps surprising."

There have been similarly puzzling results in a couple of other cases: A PET (positron-emission tomography) scan of an autistic savant taken while he was calendar calculating revealed left-hemisphere activation. An fMRI of George Widener, a savant calendar calculator and artist, showed heavily concentrated left-hemisphere activity and "sparse" right-hemisphere activity when he was calendar calculating.

These findings stand out among the large number of cases in which savant skills emerged following a left-brain injury. But they don't necessarily mean that the left-brain injury, right-brain compensation theory is incorrect, at least not in all cases.

Perhaps not all savant skills are the product of the same underlying mechanisms. The savants who demonstrated left-hemisphere activation were all calendar calculating or working on mathematical problems during their brain scans. Perhaps some calculating savants rely on different areas of the brain than savants who are skilled in art, music, or other fields. That could mean that the external similarities among savants conceal cognitive differences between those with different specialties.

Do child prodigies, too, have notable underlying differences? As demonstrated by the story of Autumn de Forest, it seems that they do.

Chapter 9

Lightning in a Bottle

As a toddler, Autumn de Forest occasionally drew pictures with crayons and markers. There were several prominent artists in the extended family, but her parents took little notice of Autumn's "little kid drawings." Her parents weren't looking for signs of artistic genius; they assumed that Autumn, who had great rhythm and a great sense of pitch, was destined for a career in music, like her father, Doug.

Autumn's parents bought her a drum set for Christmas when she was three, about a year after the family had moved from Los Angeles to Las Vegas. She showed some interest; Autumn and Doug often messed around with the drums, mocking up songs and working on rhythm. For Autumn, it was enjoyable but not life changing. Same story when her mother, Katherine, later signed her up for piano lessons: it was fun; it was play.

When Autumn was in preschool, she brought home a painting of an elephant that Doug and Katherine would later recognize as the first sign of her artistic talent. She had dashed off an array of brushstrokes that seemed to capture the elephant in motion, walking straight toward the viewer, its tail swishing behind. The somewhat abstract piece (Autumn describes it as "elephanty") seemed deliberate and artistic in a way that Doug and Katherine didn't expect from a four-year-old. But the painting got stuck in a pile of papers, lost in the rush of unpacking Autumn's knapsack. Doug didn't find it until months later.

The next flash of talent did not go unnoticed. When she was five,

Autumn went out to the garage, where her father was staining furniture. She asked to paint, and Doug gave her a stray piece of plywood, a large paintbrush, and some stain. He returned to his patina, and the two worked back-to-back. When Doug checked on Autumn, she had created two rectangles on the plywood; she had emphasized the space between them by using a darker color along the rectangles' edges. To Doug, it seemed simple yet profound. "Like a Rothko," he later told the *Tampa Bay Times*.

Autumn was nonplussed. The word "Rothko" meant nothing to her. But the feeling she got while painting, *that* was sharp and clear. "It was almost this feeling of—this feeling of contentment," Autumn recalled. "I never had that feeling before, ever, in my entire life. I still get that every time I paint. I never had it for anything else, not music, not piano, and I liked all those, but it was just—it was different."

Doug had been preparing to introduce Autumn to music composition software. But after seeing her *Plywood Rothko,* he chucked those plans. Katherine and Doug ran out and bought a few canvases. Doug rigged a small table into a makeshift artist's easel.

Autumn set to work in the garage. She painted every day after kindergarten, spending a few days on each piece. As she finished her first canvas, *Equator,* a blue abstract with a thick streak of purple slashed across it, she noticed a bag of cement spewing dust in the garage. Struck by the urge to make her work three-dimensional, she threw a pinch of cement dust on the painting. Two more small pieces followed, both abstract works. She applied a handful of cement to *Desert,* a sunset-colored mesh of orange, red, and pink; with *Pinkie,* a work with a softer mix of colors, she covered the entire canvas with cement before she applied the paint.

After finishing those three small canvases, Autumn had an overwhelming sense of wanting *more.* "I wanted a bigger space to put my ideas on," she recalled. "More color, more three dimension, more cement, more big brushstrokes, more imagination."

Her parents ran with it. Though not visual artists, both are creative types. Doug is a dark-haired musician with a smattering of facial hair who composes and plays drums. Katherine is a former actress and model who has posed for romance novel covers and had a string of roles in 1990s movies and TV shows. They had no intention of taking a wait-and-see approach to their daughter's talent. As Doug once told *USA Today,* "You see a spark and you want to do everything you can to create a wind." Doug made room in his music studio, a thousand-square-foot structure behind their house, for Autumn to work. It turned out to be a temporary arrangement. Soon, the studio was Autumn's; Doug moved his work space into the house. Katherine bought Autumn supplies. She brought home a haul of top-shelf equipment: paintbrushes, acrylics, oil paints, and, most important, large canvases.

Autumn took a trial-and-error approach to the logistics of painting a five-foot-tall canvas. If the canvas was propped up, much of the surface was out of reach for her six-year-old hands. She tried standing on buckets, balancing on both feet while she reached up with her brush. She laid the canvases on the floor but couldn't access the midsections. Eventually, Doug built a wooden bridge that arched over the canvas when it was lying flat. Autumn could sit on the bridge and tackle any part of her painting from above.

While working on one of these first large-scale canvases, an abstract piece titled *Greenie,* Autumn found her "white room." "It's this place in my imagination that I can go to, and it's a room. It's all white, and there's not a door, nothing. The only thing that's there is me, my artwork, and my tools. And that's it. I just go to this place where there are no distractions, and I feel as though I'm just engulfed in my painting," Autumn said.

She can stay there for hours. It's a place of extreme focus and contentment; Autumn celebrated it in a later painting, *Dripping Ideas,* in which the sky, her imagination, drips ideas down to the ground where they land, sprout, and bloom into paintings. Her white room is just over the hills in the background of the piece. "Sometimes I wish I could live in my white room," she once wrote. After Autumn discovered her white room, she quit piano lessons. She ran out to her art studio as often as possible.

She never took formal art classes; "I think it was because I wanted it to be from me, and me only." Instead, she learned from the masters in a self-assigned, self-designed independent study. During a visit to a bookstore, Autumn found a large tome filled with images of the American artist Georgia O'Keeffe's works. She asked for it, and her parents bought it.

Girl and book were inseparable. Autumn looked at it in the morning, thought about it during school, and rushed right back to it when she got home. She couldn't read much of the text, but the pictures entranced her. She stared at the images, soaking in the work of a professional artist.

One volume followed another: Autumn lost herself in books on Salvador Dalí, Vincent van Gogh, Andy Warhol, and Jasper Johns. Doug put together a mini-library under the TV where Autumn could keep her books.

In the studio, Autumn experimented, testing out unconventional tools. She plucked a plant out of the ground and whacked the canvas with it to create *Fire*, another large abstract. Once, frustrated by her paintbrush and "such in a hurry with my ideas," she swiped a spatula from a kitchen drawer and used it to apply paint to canvas.

She thought up a series of new techniques and worked them into her paintings whenever inspiration struck: the stink eye (hold a stick six inches from the canvas, place the palette knife between the stick and the canvas, close one eye so that the two line up, and paint the tree—"it came out pretty perfectly"); pole painting (put complementary colors on one end of the painting, place a long, thick wire just before the paint, and pull it over the paint to make "a beautiful streak of colors"); and wind painting ("how nature would paint": use a shop vac or an air compressor to blow paint across the canvas).

Doug devoted himself to keeping Autumn's materials at the ready. When he wasn't working, he prepped her canvases, videoed her in action, and soaped her up and sprayed her down when she was finished. "It takes an incredible amount of, I hate to say effort, to make things effortless for her, and I was committed to making things effortless for her," Doug said. "It was about, quite honestly, doing whatever I had to do to capture that lightning in a bottle because I didn't want the spell to break."

A constant, heavy flow of artwork emerged from Autumn's studio. The completed works migrated into the house. Autumn churned out new pieces faster than Katherine and Doug could hang them.

⤏

As Autumn approached her seventh birthday, she told her parents that she wanted to show her work. That April, Doug rented a van and packed up fifteen of her paintings. Doug, Katherine, Autumn, and Sarah (Autumn's pink Pegasus stuffed animal) drove to the Boulder City Fine Arts Festival. When they arrived, Doug wrestled their tent into place. They hauled Autumn's artwork into their tent and then waited.

The first people to filter in walked away without really engaging. "We didn't get that people wouldn't know she was the artist," Doug said. "They saw a forty-year-old man and a little girl, and they thought it was bring-your-kid-to-work day."

Autumn and Doug invented a game: Autumn pretended that the people at the event were gazelles (or, in Autumn's words, "fancy deer") and she was a cheetah; her job was to catch them. She stuck out her hand fearlessly to anyone in the vicinity, introduced herself, and launched into a discussion of her art before they could slip away. For Autumn, *talking* about her work felt like a second awakening. "I got that kind of comforting feeling," Autumn recalled, describing the sensation as similar, though not precisely the same, as the one she gets when painting. She received a badge for participating in the event; that night, she slept with it still pinned to her shirt.

The next day, they returned with a banner and a picture of Autumn, inviting people to meet the artist. People came in; Autumn, now with a day of experience under her belt, lit into her presentation. "By the time we went home on that Sunday afternoon, she was *literally* a different person than the little girl that we went down on the Friday afternoon with. I mean chemically, in her brain, she was a different person, her identity as a creator, as an artist, as an individual, as someone with things to say, with insights and legitimate statements of creativity. I mean, it was uncanny," Doug said. "It was almost like the die was cast."

The next month, Autumn brought her work to the Boulder City Spring Jamboree. There were more gazelles to snare; there was more talking about her artwork. This time, the family was prepared. They had a sign made that read, "Autumn at six." Autumn wore a pink T-shirt and perched a pink sun hat over her long red hair; she positioned herself inside the tent and waited. As the temperature rose and Autumn's face grew splotchy from the heat, her mother tried to persuade her to go inside. Autumn refused. They compromised on a lunch break, but the whole time Autumn was itching to get back— gazelles were walking right past her tent; they were getting away! Autumn won best in show for *Sunrise,* a large piece with yellow streaks of paint bursting forth from a pale yellow center. She left the event exhilarated. She couldn't imagine doing anything else with her life. As Autumn told *Girls' Life* magazine, "My spirit was on fire."

↶

In the midst of this first run of art shows—a six-month period that eventually included a trip to Malibu and a return to Boulder City— Doug decided they needed help. Autumn's artwork was selling. She sold *Paradise,* a pastel with two birds, for over $1,000; someone paid more than $1,000 for a work Autumn had donated for a charity auction; a middle-aged man offered $10,000 for *Sunrise,* though Autumn had refused to part with it. Doug researched art agents, came across a man who specialized in child artists, and contacted him. The two met at an art auction the October Autumn turned eight, and after seeing a sampling of Autumn's work, the agency took her on as a client.

At the time, Autumn's work was evolving. The images from her art books had been percolating, and she was shifting away from abstract art in favor of injecting her personality—adding a shot of joyful, mischievous pixie dust—into other artists' iconic pieces. *Dripping Ideas,* the portrayal of her creative process, was inspired by Dalí's melting clocks. When she received a Barbie from her grandmother, she decided to put her own spin on Andy Warhol's Marilyn Monroe paintings. Not wanting to dirty her new toy, she popped the head off her doll and used it as the model for *Barbie Marilyn* and *Gold Barbie Marilyn.* She followed these up with *Heart Target,* replacing Jasper Johns's bull's-eye targets with nesting hearts, and *American Graphic,* in which she substituted a pink crayon for the pitchfork in Grant Wood's iconic *American Gothic.*

When Autumn was eight, her work was sold as part of an art auction in California. Her paintings generated more than $100,000 in sales; one piece sold for $25,000. "We were definitely overcome, and we were definitely just on cloud nine, and it was unbelievable," Doug said. "Had she sold nothing, it still would've been an incredible victory—an *incredible* victory. So, it wasn't about the sales. The sales were just gravy, but, wow, that's a lot of gravy. It was proof of concept in a wonderful way." She had her first solo gallery exhibition in Stone Harbor, New Jersey, when she was ten years old.

The media caught wind of Autumn, whose mix of spunk and innocence can make her seem like a real-life American Girl character. Everyone wanted a few minutes with the "pint-sized Picasso" (*Inside Edition*), the "pocket-sized genius" (the *Daily Mail*), the "mini-Pollock" (*The Huffington Post*). She was an interviewer's dream; articulate and composed, she seemed more self-assured adult than pre-tween: "an old lady in a young person's body." She was also spunky ("a pistol," as Matt Lauer put it), charming (when asked about earnings, she quips that she's not the money girl), and so clearly *young* (hair in braided pigtails and sparkling pink headbands).

There was, of course, the usual din from the doubters: those who criticized Autumn's artwork or, attacking from the opposite perspective, those who doubted that she did it herself. The de Forests occasionally responded, and Doug posted videos online of Autumn at work.

But Autumn kept painting, and the opportunities kept rolling in. She created a series of Disney princess paintings to promote the uDraw GameTablet, signed autographs at the National Art Education Association convention, and produced the artwork used on the inside of the cover booklet for *Let Us In: Americana,* a Paul McCartney tribute album.

She had other hobbies. She loved reading, old movies, horseback riding, and animals. But painting was her unrivaled passion. She painted every day—three or four hours on school days and more on weekends. She described herself as itching to paint, compared painting to breathing, and said that she needed to create art to be happy and live. In one interview, she described the dreams she has about what the world would be like without art: it's beige, white, and boring, and it breaks her heart.

⟶

Joanne flew to Las Vegas in October 2012 to meet Autumn, who was just about to embark on a series of gallery exhibitions and auctions. She gave Autumn the same IQ test she had administered to the other prodigies. She again found that linchpin of prodigy—an exceptional working memory. But there was an interesting wrinkle: Autumn's score on the visual-spatial subtest of the Stanford-Binet was lower than her scores on the other parts of the test.

The same thing had happened once before. Lauren Voiers—the only other art prodigy Joanne had tested up to that point—had a similar dip on this part of the test. At the time, Joanne had assumed that Lauren had just run out of energy. After all, Lauren was an artist.

How could she not have excellent visual-spatial skills? But once Joanne tested Autumn, there were two prodigies, both artists, with relatively low visual-spatial scores. None of the other prodigies—not the musicians, not the scientists—showed the same pattern.

Joanne was intrigued. She had long focused on the similarities between the prodigies. After all, they shared something stunningly unique—a flash of talent so bright it created a personal spotlight. Could there really be important underlying differences between them?

⌐

Joanne contacted and tested more art prodigies; she worked with more music and math prodigies along the way. By early 2013, she had broadened her sample of IQ-tested prodigies to eighteen: eight music prodigies, five math prodigies, and five art prodigies.

The prodigies' average overall IQ score was 126, well above 100, the average score for the test. Working memory remained the stalwart of prodigiousness: the group of eighteen had an average score of 140—a score more than two standard deviations above the mean. There was only one real working memory outlier, a child with an interest in engineering who scored a 109.

But it was the differences in the prodigies' scores that were most intriguing. Whether you looked at the prodigies' overall IQ scores or sliced them up by subtest, there were variations in the art, music, and math prodigies' results. Even with a sample size of just eighteen (again, small for most studies, large by prodigy standards), many of those differences reached statistical significance—the academic equivalent of screaming for attention.

The math prodigies had an average overall IQ score of 140, well above the music prodigies' and art prodigies' group averages. Even the lowest-scoring math prodigy pocketed an overall score of 134—placing him more than two standard deviations above the general population average. The math prodigies built their way to the top of the IQ pyramid by outscoring their prodigy peers on several of the subtests. They had the highest fluid reasoning (the ability to reason through unfamiliar problems) and knowledge scores of the bunch. Same story with quantitative reasoning; the math prodigies outscored both the music and the art prodigies. The math prodigies' results were perhaps not entirely surprising. David Feldman and Martha Morelock had previously predicted that prodigious skill in science or math might require high IQ scores. This study provided the first concrete evidence that this was so.

In a group chock-full of kids with astounding working memories, the musicians out-recalled them all. They had an average score of 148 for this subtest—more than *three* standard deviations above the mean for the general population.

But it was the visual-spatial breakdown—the subtest that had launched Joanne's interest in the differences in the prodigies by specialty in the first place—that raised the most questions. The artists' dip in visual-spatial scores held. As a group, the art prodigies had an average visual-spatial score of 88, their lowest mark on any part of the test. Not only did they score significantly lower than the math and music prodigies, but the artists tended to score below the average for the general population on this subtest. It almost seemed like a deficit in visual-spatial abilities was *necessary* for artistic talent.

It was counterintuitive: Shouldn't artistic composition require excellent visual-spatial skills? The answer is yes—sort of. Artists do rely on visualization, the process of "seeing with the mind's eye" or conjuring an internal representation of an object, landscape, or event. The nitty-gritty of it comes down to the *type* of visualization upon which they rely: *object visualization* versus *spatial visualization*. An *object visualizer's* mental imagery is detailed and focused on physical attributes—an item's color, shape, brightness, and size. A *spatial visualizer's* mental imagery is more three-dimensional; it's focused on an object's position relative to its surroundings, its movement through space, and its physical transformation.

A 1985 study involving two brain-damaged patients illustrates the difference: Following a car accident, Patient 1's object imagery declined (he struggled to recognize people either in person or in photographs). His spatial imagery, though, remained intact (he could give detailed directions and locate major cities and states on a map). Patient 2's object imagery survived a brain hemorrhage (he could still identify objects, animals, and people). His spatial imagery didn't fare so well (he often collided with objects in his path and couldn't distinguish left from right).

So which type of visualization was the IQ test measuring—object or spatial?

The Stanford-Binet, like many intelligence tests, is rooted in the Cattell-Horn-Carroll (C-H-C) theory of cognitive abilities. The visual-spatial subtest is intended to measure the C-H-C's visual processing ability—an individual's ability to recognize an object, understand its location, and predict its motion or transformation. Its "core" is the "ability to perceive complex patterns and mentally simulate how they might look when transformed (e.g., rotated, changed in size, partially obscured)."

In other words, the visual-spatial subtest measures *spatial* visualization. This type of visualization is often linked to math and science talent. Consider, for example, that numbers appear to the math prodigy Jacob Barnett as shapes that he can manipulate to solve problems. Galileo similarly visualized motion through space to develop the idea that objects in a vacuum fall at the same rate, regardless of weight. Both of these problem-solving techniques rely on spatial visualization.

Not surprisingly, then, the math prodigies walloped the visual-spatial subtest. They had the highest average score, by far, of the three groups: 142 compared with 117 for the musicians and 88 for the artists. Even the lowest-scoring math prodigy snared a 132 on the visual-spatial subtest—a score more than two standard deviations above the average for the general population.

In theory, this left object visualization, a skill tied to artistic ability, unmeasured. But were the artists' object visualization abilities a complete unknown? Perhaps not.

The psychologist Maria Kozhevnikov and her colleagues, the team behind some of the most interesting research on distinguishing object visualization from spatial visualization, have hit on an intriguing find that may provide a clue as to the artists' object visualization abilities. It turns out that not everyone is a visualizer. Some people—verbalizers—approach problems wielding words and logic instead of the visualizers' mental imagery. But Kozhevnikov and her colleagues have found that people who *are* visualizers experience a trade-off between object and spatial visualization—the price of excelling at one is a deficit in the other. Scientists' above-average spatial visualization generally comes hand in hand with below-average object visualization. Visual artists' below-average spatial visualization is generally accompanied by above-average object visualization.

Viewed from this perspective, for an artist, a low visual-spatial score isn't a deficit. It's a badge of honor. It marks a different type of talent, one not directly measured by the Stanford-Binet; it hints at a reservoir of vivid, detailed object imagery—a critical attribute of an artist.

―

Most of Joanne's work had been geared toward unearthing the inner workings of the prodigy mind. She had long assumed that she would find core similarities in the group. After all, the flash of childhood achievement they shared was extraordinary; surely that rare explosion of talent, when it occurred, had to be caused by the same underlying mechanism.

To some extent, that assumption was borne out by the prodigies' extraordinary memories and excellent attention to detail. But the prodigies' seemingly irresistible attraction to their fields of interest suggests differences between them. Autumn de Forest's road to painting, for example, feels like destiny, like something she was meant to find. No one pushed it on her. Her parents introduced her to drums; they provided piano lessons. Autumn was a willing-enough student. She had fun experimenting with music. But there was no spark. Music didn't speak to her. As Autumn put it, "I hadn't really found it yet; I hadn't really discovered it yet." It's almost as though art were *waiting* for her. Once she found it, the connection was electric. Autumn latched onto painting as if it were a drug, a tonic, and a path to survival.

Most prodigies tell the same story. They didn't waffle over which activity to pursue or weigh the pros and cons of different endeavors. They felt *called* to something specific. They discovered that something early and clung to it with both fists. Around the time he was two, the music prodigy Jay Greenberg began drawing cellos even though, as far as his parents could remember, he had never seen one. He requested one of his own. Music sprouted from his mind involuntarily. "I just hear it as if it were a, a smooth performance of a work already written, when it isn't," he said during a *60 Minutes* interview. Jacob Barnett, the science prodigy who took a college astronomy course at eight years old, was obsessed with light and shadows as a baby and, at three, grew so engrossed with an astronomy book— a book so large he had to drag it around on the floor—that his dad had to duct tape its spine to hold it together. The music prodigy Jonathan Russell began saying the word "violin" at eighteen months and repeated the word every time he saw one. Soon thereafter, he could pick the sound of the violin out of audio recordings.

The prodigies' IQ-test results suggest that the reasons each chose a particular specialty had to do with deep-rooted differences between them. A quick glance at their visual-spatial scores reveals that something is going on in the prodigies' brains that sharply divides the math from the art specialists—a critical *cognitive difference* that led some prodigies to excel at art and others at math.

The object and spatial visualization abilities of the art and math prodigies, moreover, aren't only distinct in practice; they rely on separate neural pathways: the ventral pathway (the "what" pathway) processes object imagery, while the dorsal pathway (the "where" pathway) processes spatial imagery. The art prodigies seem to have an express pass to the ventral pathway, while the math prodigies have the same sort of access to the dorsal pathway.

Autumn then isn't just a kid with a souped-up memory, a turbo-charged engine that would have driven her to blindly latch onto trombone, paintbrush, or equation with equal fervor. She's a precision instrument. She always had the horsepower for prodigiousness, but she had to find the right lock-and-key fit to unleash it. She needed a task that called upon her very specific mental profile to reveal her extraordinary abilities. She isn't just *a prodigy;* she's an *art prodigy.*

But if there are important distinctions among the prodigies, does it make sense to think of them as belonging to the same group? Once you dig into the children's cognitive profiles and come up with important differences, is there really even such a thing as a "prodigy"?

There's another field of research in which researchers face the same question, another field of research in which a focus on behaviors has obscured underlying differences: autism. Is it really one distinct condition? Over several decades, researchers gritted their teeth and tried to find a reliable way to distinguish autism from other disorders. They ransacked autists' behaviors, cognitive tendencies, and, eventually, genes. But despite the mighty efforts of a slew of scientists, those diagnosed with autism defy neat packaging. If a unifying factor that cleanly separates autists from non-autists exists, it's an elusive beast in a field brimming with hunters.

The difficulty of isolating autism-specific symptoms begins at the behavioral level. Kanner's original 1943 paper described a number of behavioral traits that he had observed among his autistic patients, and he specified that the condition's defining characteristic was "the children's *inability to relate themselves* in the ordinary way to people and situations from the beginning of life." But in the early years of autism research, scientists attempting to investigate this new condition used varying criteria. By the early 1970s, one researcher had already observed that there was no symptom that definitively indicated autism; the symptoms associated with autism were also characteristic of a number of other disorders. In a classic 1979 paper aimed at clarifying the clinical picture of autism, the British researchers Lorna Wing and Judith Gould concluded that their results called "into question the usefulness of regarding childhood autism as a specific condition." Recently, a team of researchers proposed that the behavioral symptoms used to diagnose autism are largely independent of one another. Even autistic *siblings* often have significant differences in their social and communication abilities. This variety in behavioral symptoms is well captured by the oft-repeated quotation that "if you've met one person with autism, you've met one person with autism."

Digging a level deeper didn't help. Researchers struggled mightily to identify a core cognitive characteristic of autism—an underlying trait that would explain all of autism's symptoms. One by one, the first generation of cognitive theories—theory of mind, weak central coherence, and executive dysfunction—were revised or abandoned. Their modern counterparts and new cognitive theories seem to be a better fit, but no one has yet identified a cognitive trait or traits possessed by all autists and not possessed by any non-autists.

It's the same story with genetics. Initial optimism that scientists would isolate common genetic risk factors for autism quickly faded. Researchers found not one but dozens of genes—some studies predict that the ultimate count will be in the hundreds—implicated in at least some cases of autism. Even the most prevalent of these genetic variants are tied to less than 1 percent of autism cases. In 2009, one team of researchers observed that "the genetic architecture of autism is as exquisitely complex as is its clinical phenotype" (basically, there are as many autism-related genes as there are variations in individual autists—many). It turns out that even autistic siblings often have different genetic risk factors for autism. Instead of a single genetic pathway to autism, it appears there are many.

This heterogeneity of behaviors, cognitive traits, and genetics has led many researchers to argue that the notion of autism as a disorder with a single underlying cause must be abandoned. The prodigies, too, have distinct cognitive traits: different strengths undergird the abilities of the artists, the musicians, and the scientists. Do they also differ from one another genetically, according to either specialty or another unidentified variable? Could there be different genes that result in prodigious talent, just as there seem to be many genetic pathways to autism?

⟶

Guy Rouleau, an affable, gray-haired French Canadian, is the director of the Neuro (McGill University's neuroscience research hospital), a professor at the University of Montreal's medical school, and a genetics rock star. Over the course of his career, he has tracked down a number of disease-causing genes. He specializes in de novo mutations, the genetic mutations that are present in an individual but not in either parent. He and his team have tied such de novo mutations to both schizophrenia and autism.

All of the mutations he has identified have led to a decrease in functionality; they often negatively impact an individual's quality of life or daily living skills. But, in theory, if de novo mutations can result in a decrease in functionality, they can also lead to enhanced abilities. It was this idea that inspired Rouleau's long-burning interest in trying to track down a savant gene, the name he used for a yet-unidentified mutation that he thought was likely to enhance an individual's cognitive functioning and result in savant-level skills.

Others had tried to find such a gene before. One group of researchers attempted to cut through some of the complexity of autism genetics by focusing on autists with a savant skill and their family members, and they found some evidence of a connection between such savant skills and chromosome 15. But when another team tried to replicate those results, they failed.

Rouleau wanted to see if he could get better results by looking for de novo mutations. He and his team tried to track down savants willing to provide DNA samples, but they didn't get terribly far. Savants are rare; there are often gatekeepers—friends or family or managers—unconvinced that participating in research is in the individual's best interest. For a while, the project stalled.

One of Rouleau's collaborators read about Joanne's research. There was probably some overlap between prodigies and savants, the team thought, given their shared abilities—why not see if they could get something going on the prodigy front? A few months after Joanne met Autumn in Las Vegas, she had breakfast with Guy in Montreal, and the two discussed a potential collaboration.

Soon after, Joanne and her Ohio State collaborators would find evidence of a possible inherited link between autism and prodigy. But what if Guy's theory about de novo mutations is right, too: What if, just as there's more than one genetic path to autism, there's also more than one genetic path to prodigy? Is it possible that prodigy, like autism, isn't a single condition but a multitude of related conditions, all grouped together because of the similarity of their symptoms? Does prodigy, like autism, have multiple different genetic origins—important underlying variations hidden by a mask of similar behaviors? After nearly two years of discussions, Joanne sent Guy and his team the prodigy DNA. They are now searching for de novo mutations that could provide another piece of the prodigy puzzle.

Chapter 10

The Recovery Enigma

Lucie is still wondering: Did her sons recover from autism?

To her, it certainly seems like it.

Her older son, Alex, has long since had his autism diagnosis stripped away, and he's humming through the mainstream school system. He's an excellent swimmer, a skilled Lego architect (his creations include a four-foot replica of the *Queen Mary 2*), and an outgoing chatterbox. He's popular and enjoys social situations: one morning over breakfast, the child who used to retreat to his room when company came over asked his parents to have more parties. By the time he was ten, his school behavior reports weren't just passable; they were excellent.

Her second son, William, is also becoming more socially adept and a better communicator. He was never formally reassessed regarding his autism diagnosis—he's not in therapy, so the government hasn't requested another evaluation, and Lucie and Mike don't care about the label, so they've never bothered—but every year he seems happier and more engaged. He's more open to new experiences, participates in group projects at school, and, more and more often, joins the other kids' games on the playground.

His mind is still insatiable. He's enrolled in Stanford's online high school math courses and draws Fibonacci spirals in chalk. One Thanksgiving, when his parents gave thanks for family, jobs, Canada, and health, and Alex gave thanks for the family Lego room, William, then seven, expressed his gratitude for irrational numbers.

To Lucie, the boys' autism seems to have disappeared; it looks an awful lot to her as though her sons have recovered.

In scientific circles, though, "recovery"—the notion that an individual no longer has any symptoms of autism—was, for a long time, something of a dirty word. In the earliest autism studies, there was the occasional mention of a former patient who no longer seemed autistic, but the scientists who described such rare cases never used the word "recovered." More common were reports of individuals who adjusted somewhat to school and their community but who didn't seem to have fully shed autism's symptoms. These individuals might have progressed through school, for example, but they still generally had limited communication and obsessive tendencies. As the author of a 1956 study put it, even those with the best outcomes often remained "somewhat odd."

The word "normal" began popping up in autism studies in the late 1960s. It wasn't used often. It applied to a kid here, a kid there. One study described a child who developed normal friendships; there was a report of an adult who lived independently and had a job and friends. But some scientists were still skeptical that these types of cases represented true recovery. Many thought autism was a lifelong condition; recovery just wasn't a possible outcome.

That attitude has only recently begun to change, and that change is largely due to the work of the University of Connecticut professor Deborah Fein. She and her team have tackled the recovery question head-on by investigating "optimal outcome" cases—individuals who were diagnosed with autism but who *seem* to have recovered.

The team put these optimal outcome kids through a series of tests to see whether any signs of their autism remained. What they found sent a small earthquake through the autism community: The optimal outcome kids didn't differ from their traditionally developing counterparts in communication skills or social skills. They attended mainstream schools and had developed friendships with their typically developing peers. They didn't have any residual academic difficulties.

As far as the examiner's eye could see, all signs of autism were gone.

Still, the researchers shied away from actually using the word "recovery." They cautioned that an optimal outcome isn't necessarily a possibility for every child and that it shouldn't be considered the only good outcome for children diagnosed with autism.

But Sally Ozonoff, the joint editor of one of the journals to publish the team's findings and herself a major autism researcher, declared it time to begin taking "the 'r' word" seriously—and researching the idea rigorously.

For the first time, it appeared scientifically validated that, with respect to recovery, perhaps not all hope is false hope.

⟶

If recovery is real, why—and when—does it happen?

This is still largely a mystery. The uncertainty surrounding any individual autist's prognosis highlights the urgency of unraveling autism's heterogeneous underpinnings. Despite presenting with similar symptoms, individuals with different genetic profiles or environmental exposures (and any combination thereof) will likely respond differently to different types of treatment (and some might even recover or make significant gains in functioning regardless of the type of treatment they receive). So long as everyone who presents with the collection of behaviors we call autism is grouped together, there's no way to know in advance which course of treatment will be most effective for any particular person. As the former NIMH director Thomas Insel once put it, "Symptoms alone rarely indicate the best choice of treatment."

Lucie, for example, credits behavioral therapy—an intervention designed to change behaviors through learning-based techniques—for her older son's language and social gains. When she first brought Alex to the Portia Learning Centre where he received his therapy, he was nonverbal and withdrawn; within a few months, he was speaking and growing more interactive.

Behavioral therapy was first tested as a potential autism treatment in the 1960s, a time when there was a growing consensus that other types of autism therapies hadn't worked. In one of the earliest experiments, a team of researchers attempted to teach autistic children to press a button or deposit a coin at a particular time (the desired behavior); if they succeeded, they were rewarded with candy, coins, music, or games (the positive reinforcement). The autistic children learned the encouraged behaviors more slowly than the controls, *but they did learn them.* Two years later, another team reported that they had used behavioral therapy to reduce an autistic preschooler's tantrums and improve his sleep and communication. In contrast to the failure of other therapies, behavioral therapy seemed to be brimming with potential.

It got a dramatic boost from a series of experiments conducted by the UCLA psychologist O. Ivar Lovaas, a man whose team infamously combined positive reinforcements, like food and expressions of approval, to build communication and social behavior, with "aversive stimulations," including slapping the children and administering electric shocks, to stymie undesirable behaviors. In the early 1970s, the researchers reported that every child in the study showed some behavioral improvement. In a follow-up investigation published in 1987, Lovaas claimed that nine of the children (or 47 percent of the experimental group) receiving more than forty hours of behavioral therapy each week finished mainstream first grade and achieved an average-or-above IQ score—"normal functioning," as he described it. A few years later, eight out of nine of these children were attending traditional schools and blending in with their peers.

Such results naturally attracted a great deal of scrutiny (even apart from any concerns over the use of aversives). Some scientists suggested that Lovaas and his colleagues had limited their experimental group to autists with relatively better prognoses; others were skeptical of whether the bar for "normal functioning" had been set sufficiently high.

Since then, behavioral therapy has been explored from many angles. Some studies tinker with the protocol; others test the importance of the specific type of therapy, the number of hours administered, whether the therapy was led by a professional or a parent, and the age at which the child started therapy.

Despite this growing body of research, the advisability and efficacy of behavioral therapy remains somewhat controversial. Behavioral therapy's critics view some of its aims as misguided. They view efforts to eradicate autistic behaviors like hand flapping, for example, as interference with autistic self-expression; such efforts, this line of thinking goes, have little to do with autists' long-term quality of life. Parents and teachers should try to understand the root cause of these behaviors rather than try to make autistic communication resemble neurotypical communication. As Laurent Mottron, a prominent autism researcher and professor at the University of Montreal, emphasized, the goal shouldn't be to make autists not autistic; it should be to recognize the unique contributions of autists and the distinct ways that autists process information. School and work environments should then be adapted to accommodate their needs.

In terms of effectiveness, a 2014 meta-analysis commissioned by the United States government concluded that there was "moderate" evidence that early intensive behavioral intervention improved language and cognitive abilities and "low" evidence that such therapy improved adaptive behaviors, social skills, or the severity of core autism symptoms. Still, a 2015 review concluded that early intensive behavioral intervention is the autism treatment "with the greatest amount of empirical support." The U.S. surgeon general and Autism Speaks, a prominent advocacy and science organization, both support some types of behavioral therapy. Behavioral therapy (still built around principles of learning but now offered in a number of different forms) has become a mainstay of autism treatment in the United States.

Behavioral therapy may help some autists, but it's certainly not a cure-all. Not every kid who receives behavioral therapy shows significant gains in communication and social skills, and some kids show drastic improvement in these areas without it. In Fein's optimal outcome group, for example, the kids who recovered from autism were more likely to have received behavioral therapy than other children. But there were still optimal outcome kids who had received relatively little early intervention, and there were children who did not recover even though they received intensive early intervention.

While Lucie's son Alex, for example, seems to have responded extraordinarily well to behavioral therapy, it was less successful for William. Lucie believes that William's constantly changing interests made it difficult to identify compelling positive reinforcements, an important component of behavioral therapy. In his case, medication eased his social anxiety and perfectionist tendencies. For Jacob Barnett, it was engaging more and more deeply with topics about which he was passionate that seemed to build his social and communication skills—a burgeoning approach to autism known as training the talent.

~

Almost from birth, Ping Lian Yeak was different from his older sisters. He didn't babble when his sisters had. He didn't say his first word until he was two years old. Even then, there was only the occasional utterance; his speech was nowhere near where his older sisters' had been. At three, he still barely spoke. He wasn't quiet, though. He was prone to violent, screaming tantrums; he kicked and bit his parents and sisters. He hardly ever slept, and he didn't like to be hugged or touched.

As a toddler, he was hyperactive. If his parents lost their grip on him, he'd bolt. Once, after slipping out of their grasp at a mall near their home in Kuala Lumpur, Ping Lian ran to the food court and drank out of someone else's drink. His parents, Sarah Lee and Min Seng, sent him to a Montessori school; Ping Lian tried to escape by climbing over the front gate.

Observing—really more than observing, *studying*—interesting scenery and images was the only activity that calmed him. Whenever the family passed a bookstore, Ping Lian darted inside and tracked down the house-and-garden section. He gathered home design, architecture, and landscaping magazines and carefully paged through them, scrutinizing the pictures. He did the same with outdoor scenery. When the family went to the lake, Ping Lian ran to the edge of the water, where he would sit peacefully for long periods of time, intently studying his surroundings.

At first, his parents thought it was just a phase. Ping Lian, who had a round face, dark eyes, and a sweet smile, was handsome. His gross motor skills were good. Other parents reassured them that boys just developed differently than girls did. They were more active; they spoke later.

But when, at nearly four years old, Ping Lian still spoke only a few words, his parents had him evaluated by a speech therapist. During his appointment, Ping Lian kicked the door and threw things. He was restless and rarely made eye contact. He imitated some speech sounds and occasionally pointed, but his communication was limited; the evaluator estimated that he could say only about sixty words, mostly single words and stock phrases like "I don't want."

The somewhat cryptic wording of his final report characterized his condition as follows: "Ping Lian presents with moderate to severe speech-language delay characterized by deficits in receptive and expressive language skills and reduced pragmatic (social interaction) skills. In addition he presented with attention deficit and hyperactive traits. Ping Lian's ability to comply was negligible at this time." Sarah and Min Seng took him for an evaluation at a hospital where they were told that Ping Lian had ADHD with autistic features.

Sarah, a petite woman with black, curly, shoulder-length hair and a happy, bustling energy, was relieved. She focused on the hyperactive aspect of the diagnosis, and that seemed like a common enough problem. Sarah and Min Seng signed Ping Lian up for the recommended speech and occupational therapy. They waited for him to get better.

About a year later, Ping Lian's speech therapist chastised Sarah for her lackadaisical attitude: You can't take this so lightly, the therapist told her. You need to do more for your son. Sarah and Min Seng bought a computer with an Internet connection so they could learn more about Ping Lian's condition. The more they read, the more it became obvious that Ping Lian wasn't just hyperactive; he was autistic. They were shocked and distraught. "Devastated," as Sarah put it. They thought about Ping Lian's future, and what they envisioned crushed them.

Sarah and Min Seng set out to learn everything they could about autism. They read about it online, attended seminars, joined parent support groups, and pooled their money with other families to buy autism books from overseas. They read somewhere that autists don't understand love and affection (the same idea that had led Lucie to kick one of Alex's behavioral therapists out of the house), and the thought was crushing. "I tell myself, if he can't talk, it is not a big problem, but if he has no feelings of love, that is a big problem," Sarah recalled. Sarah and Min Seng focused on Ping Lian's emotional development. They let him sleep between them and stroked him soothingly. They played him stimulating music, and, even when he was sleeping, they whispered that they loved him—anything to try to break through to him.

Ping Lian continued his speech and language intervention and occupational therapy. He left his Montessori school for a combination of homeschooling and classes at a special education center. His parents wanted him to begin behavioral therapy, but they couldn't afford a professional therapist. Instead, Sarah hired and trained a student to come to her house a few hours a day and help with Ping Lian's behavioral therapy.

Sarah pitched in at night and on the weekends. Ping Lian's fine motor skills were underdeveloped, and he still couldn't hold a pencil properly or use scissors, so she held his hand and guided him through the motion of tracing letters and numbers, pictures and shapes. It was hard for Ping Lian to complete even simple tasks, such as standing up or sitting down, but he managed to sit still for tracing. It was a small victory, but it felt like a milestone; it "at least let us feel good that we can get him to sit down and do something," Sarah recalled. Sarah spent months positioning and guiding Ping Lian's hand as he held the pencil. Eventually, Ping Lian could trace on his own. He seemed to enjoy it; some days he traced more than ten pages' worth of material.

When Ping Lian was eight, he went to get ice cream with his father. When he returned, he rushed into the house and ran upstairs. Sarah followed and found Ping Lian sitting at his table, studying his ice cream wrapper, carefully reproducing the pictures on it, "totally focused and full of energy." It didn't hit her until later: Ping Lian wasn't just tracing; he was *drawing*.

From that point on, he drew constantly. He did it independently, without any coaxing. During therapy, he drew "anywhere and everywhere," on whatever he was working on or whatever paper he had around: his books, his exercises, even his schedule. "He seemed almost obsessed with drawing," Sarah recalled in a book she is writing about raising Ping Lian.

That Christmas, the family spent some time at Sarah's sister's house with their relatives. Sarah watched as over the course of several days the rest of the cousins played, and her nine-year-old son drew—constantly. She had long feared for Ping Lian's future. What kind of career could he have? How would he spend his days? Sarah had toyed with a few possibilities: maybe Ping Lian could work at a dim sum shop or café or maybe as a cleaner or a gardener—jobs that would give him some measure of independence but where he wouldn't have to talk much. But Sarah began to wonder, could he be an artist?

She signed him up for art classes with three different teachers. None of the three had any expertise in working with autistic children. Sarah gave them articles on autism and told them that she was planning a career in art for Ping Lian. The art teachers were nervous. It's extremely difficult to make a living as an artist, they cautioned. Sarah told them that she was prepared to wait. She would give Ping Lian plenty of time to develop into a professional artist.

Ping Lian's output climbed at an almost frenetic pace. He sketched his family and movie characters; he painted animals and landmarks. There was a roughness to his early drawings, a rudimentary feel, but some of the details—the folds in Ronald McDonald's pants, the angle of Woody's body as he leans against Buzz Lightyear—hint at an observant eye and growing precision.

Sarah focused on convincing Ping Lian that he had talent, "brainwashing him," as she puts it. Every day, she told him that he was an artist. At the end of the year, Ping Lian's work was included in a group art exhibition, *Different Strokes—Diversity Through Art,* at Malaysia's National Art Gallery. For the first time, Ping Lian, then ten years old, was publicly presented as an artist.

↤

Two months later, Min Seng died of a heart attack.

Ping Lian thought the funeral was a party. He enjoyed the food; he enjoyed the people. He didn't understand what death *meant.* Sarah had to find a way to explain to him that his father wasn't coming back. Once he grasped the finality of what had happened, Ping Lian worked through his emotions by drawing pictures.

Sarah left her job to spend more time with Ping Lian, and he delved further into his artwork. He still had a short attention span for most activities, but he spent hours at a time drawing and painting, often humming while he created. His productivity was stunning; the house was soon brimming with his art. He worked quickly and with intense focus, appearing, as one reporter put it, "deep in the eye of his own creative hurricane."

Ping Lian still sketched and painted a wide variety of subjects— horses and family and flowers—but he began focusing more heavily on buildings and landmarks. He painted the Kuala Lumpur Railway Station and the Petronas Twin Towers, a pair of Kuala Lumpur skyscrapers; he painted a series of academic buildings, including the University of Malaya schools of dentistry and economics.

The drawings have an abstract quality: Buildings lean at haphazard angles. Trees are partially individual entities, partially enmeshed in the landscape. Color darts through the sky, the buildings, and the scenery in a way that makes even inanimate objects seem lively and joyful; just like Ping Lian, they hum with activity.

He exhibited his art often. By the time a year or so had passed since Ping Lian's first exhibition, he had already sold many original works and prints. One of his pieces, *Ubudiah Mosque I,* a colorful depiction of a crisp, white-and-gold Malaysian mosque, raised north of $25,000 at a charity fund-raiser.

There was some pushback from Ping Lian's teachers over the amount of time he devoted to art. They worried that he was focusing too heavily on painting and drawing at the expense of other skills.

Sarah never saw it that way. From her perspective, the opposite seemed true: helping Ping Lian develop as an artist didn't impair his development; it *contributed* to it. By working with multiple teachers, Ping Lian was interacting with more people; the more he practiced socializing, the better he got at it. His frustration and hyperactivity decreased.

Plus, drawing made him so *happy.* Sarah could see it in his actions; she could see it in his art—the forms he drew, the buoyant colors he chose. In a series of paintings titled "Happy Fishes," the vibrant orange-and-gold fish almost seem to be smiling. Even if it didn't pan out as a career, art was still a great hobby.

Instead of encouraging Ping Lian to back off his artwork, Sarah tried to expand his audience. She sent a box of Ping Lian's prints to Laurence Becker (the same educator turned agent who worked with Richard Wawro), hoping to arrange a U.S. exhibition. Laurence and Rosa C. Martinez, a woman who later founded Strokes of Genius, a nonprofit organization that develops and promotes autists' artistic skills, set up an exhibition in Brooklyn, and Ping Lian and his mother traveled to New York.

On the drive to Brooklyn from the airport, Ping Lian hit the windows; the sound of the air conditioner seemed to trouble him. At the exhibition, he reveled in the attention he received. Even in a foreign country, even surrounded by strangers, he felt a strong call to create. "Here we are in the gallery, and it didn't faze him," Rosa recalled. "He would walk right through the crowds like they weren't there, throw his pad on the floor, lie down, and start drawing—anywhere."

When Ping Lian was twelve, the family moved to Sydney. Ping Lian was captivated by the city's architecture. He did a series of intricate drawings and paintings of the Sydney Harbour Bridge and Opera House; his subjects later expanded to the animals at the zoo and the creatures at the aquarium, and he filled his pictures with fish, koala bears, and kangaroos.

Sarah introduced Ping Lian to an art market in Sydney Harbour where she and Ping Lian began making weekly appearances. They hung prints of Ping Lian's vibrant pictures in a white tent and propped them up on a table in front of the booth.

Trips to the market were partly a commercial exercise, but, more important to Sarah, they were also an extension of Ping Lian's therapy. Sarah and Ping Lian observed the bustle of the market and practiced interacting with other people. Sarah worked with Ping Lian on relevant tasks: packing his materials, loading the car, and, eventually, teaching him how to set up the stall. They practiced counting money and serving customers.

Ping Lian's reputation flourished. He was a featured artist at a pop-up art exhibition in Sydney. Two groups sponsored a permanent showcase of his work at the Art Commune in Malaysia; for several years, Ping Lian's art hung in a curved white hallway until his permanent showcase was moved to a hotel in Malaysia. In the United States, his artwork was featured in exhibitions at Carnegie Hall, 100 United Nations Plaza, and New York's Port Authority Bus Terminal. He participated in an exhibit in Tokyo and another in Singapore. His cityscapes ("imposing in their intricacy"), animal portraits ("vivid splashes of color"), and style ("bold strokes and cheerful colours") were celebrated in newspapers, magazines, books, and documentaries.

During a television interview, Sarah asked Ping Lian what he wanted to be when he grew up. Ping Lian, then an expressive seventeen-year-old, raised his hand above his head and smiled. "Great artist."

⌒

Ping Lian's communication is still limited. He speaks only a few words at a time.

But the family has seen him express emotion and demonstrate affection—qualities they once worried they would never witness—many times. They rejoiced when, while still a small child, Ping Lian rushed to soothe a crying baby at a shopping center. After his father and grandmother died, Ping Lian told his mother that he missed them—an unexpected communication of deep emotion. When he goes to sleep, he tells his mother that he loves her, and sometimes he gives her a kiss. He lets Sarah hug him.

He's less hyperactive. He uses the computer and plays the piano. He washes food, cleans dishes, hangs clothes on the clothesline, and vacuums the house. He loves traveling to new cities and staying in nice hotels. Just as Sarah hoped, Ping Lian's art helps him to express himself; it helps him find common ground with other people. It has improved his self-esteem.

From Sarah's perspective, Ping Lian's autism has had a positive impact on the rest of the family. He is on a unique journey; his talent is a gift from God. As she sees it, working with Ping Lian has molded her and her daughters into more patient and compassionate people. Life has many challenges, but as she once wrote, it has "become so meaningful and purposeful."

⌒

Training the talent isn't usually a formal therapy that comes with a professional therapist and detailed instructions for parents. In most cases, it's a somewhat ad hoc approach to autism. The basic idea is to identify autists' strengths and interests, help to develop them, and use those areas of interest to engage and teach the individual. It's an intriguing approach to autism, but it needs to be scientifically vetted for effectiveness. While behavioral therapy has been put through many, many studies, and some of the newer models of behavioral therapy (and another variety of therapy, Floortime) share some common ground with training the talent, no controlled experiments comparing training the talent with other therapies have yet been published.

The delay in exploring training the talent as a comprehensive autism treatment may stem from the fact that it involves developing autists' obsessions, which have historically been somewhat neglected by researchers. Though a tendency toward obsession is a widely recognized autistic trait (and included in autism's diagnostic criteria), these obsessions are the subject of far less research than autists' social and communication abilities. At least a few scientists, moreover, view autistic obsessions as an impediment to overall development. Some have proposed that when these obsessions take the form of a savant-like skill, there might be a trade-off between the development of that skill and the development of communication and social skills. Kanner, for example, questioned in his 1943 paper whether the pride that his autistic patients' parents took in their children's extraordinary memories—a pride that he believed led these parents to "stuff" their children with information—interfered with his patients' communication skills. More recently, scientists considering the case of a savant artist questioned whether the time he spent painting and building models—solitary pursuits—impeded his linguistic and social skills.

Like Sarah Lee, many parents adopt a train-the-talent approach to autism not at a scientist's or a therapist's recommendation but out of an intuition that allowing their children to pursue their interests may prove engaging and beneficial—or at least soothing. There's some reason to think that they may be right. There's evidence that, in contrast to individuals with obsessive-compulsive disorder, who tend to experience their obsessions negatively, autists find time spent pursuing their obsessions enriching and enjoyable. These strong interests can also serve as common ground for friendships and social connections. Several small studies have found that incorporating autists' obsessions into games increases the children's social interactions with peers and siblings; others have found that using obsession-related rewards improves autists' performance on tasks, reduces tantrums and aggression, and increases social interactions. Stories like those of Jacob Barnett and Ping Lian Yeak, moreover, turn the idea of a trade-off on its head. In both cases, developing special skills *promoted* social interaction and life-skills development.

This idea has really been around since the beginning of autism research. In 1944, in his first published paper on autism, Hans Asperger described an autistic child who was clumsy, failed to recognize close acquaintances, and ignored school subjects that didn't interest him. Despite these challenges, he was able to parlay an all-encompassing interest in math and shapes into an academic astronomy career.

In 1971, Kanner published a follow-up study in which he investigated the status of his original autistic patients. The patients whom Kanner identified as "the two real success stories" both had caregivers who intuitively trained their talents. As a child, Donald T. was withdrawn, obsessed with spinning objects, and prone to temper tantrums. A couple who took him in for several years channeled his interests in measurements and numbers: in their care, Donald dug and measured a well and learned to plow corn while counting the rows. Later in life, he secured a job as a bank teller. The second success story, Frederick, was slow to speak and had limited interest in people as a child, but his parents and the instructors at his school helped him build his social skills based on his interests in music and photography. He ultimately received vocational training running a copy machine; he, too, held a steady job.

The savant expert Darold Treffert cites training the talent as the approach adopted by the families of many of the savants with whom he has worked, as does Becker. Temple Grandin, the famous animal scientist and autism activist, urges parents to develop their autistic children's talents.

Not every child, of course, has or will develop a skill on which this technique could be used effectively. Training the talent isn't necessarily a substitute for other kinds of therapy, nor is it a mind-set meant exclusively for autists. But it's an orientation toward autism that seeks to develop the individual's capabilities and interests, and it emphasizes autists' strengths rather than their challenges.

It's still early days for training the talent. It needs to be further developed, standardized, and tested on large samples. But even once such programs are built and have been rigorously examined, investigating different therapies still only gets you halfway. It's likely that a train-the-talent approach to autism, just like behavioral therapy, will be more effective for some kids than for others. Behavioral therapy, for example, might be highly effective for those with particular types of underlying biology, but less so for those whose autistic symptoms stem from different biological roots.

The idea here isn't to endorse train the talent as an alternative to other autism therapies. It's to point out the practical importance of the ongoing basic science efforts to identify the underlying differences between autists and the specific mechanisms that can lead to autism. Eventually the findings that stem from this work may allow for more precise choices regarding intervention type and frequency. Trying to find a treatment that works for everyone based solely on similarity of symptoms is like trying to treat all people who have trouble breathing by giving them the Heimlich: it will help those who are choking, but it's probably not the best answer for a person having an allergic reaction.

To make informed decisions about treatment, you first have to know what underlying condition you are trying to treat. Until scientists parse out the different mechanisms that lead to autistic symptoms, there's no way to predict which course of treatment will work best for any individual (or to predict who will recover from autism and why). There's no way to tell, for example, who's an Alex, a child who may respond extremely well to behavioral therapy, and who's a Jacob Barnett or a Ping Lian Yeak, a child who may respond much more dramatically to efforts to support and develop his interests. The need to make optimal treatment decisions is one more—very pressing—reason for research to focus not just on autism's behavioral symptoms but also on the genetic and biochemical abnormalities that lie beneath. As Insel put it in a 2014 blog post about autism, "The best way to better services will be through better science."

The possibility that there may be a genetic link between prodigy and autism then suggests an interesting question: Could the way to better science be paved by child prodigies?

Chapter 11

The Next Quest

Child prodigies have long been a riddle, their abilities a great un-answered question. David Feldman and Martha Morelock once complained—only somewhat facetiously—that "divine inspiration, reincarnation, or magical incantation" were the best explanations for child prodigies that science had to offer.

From the day Joanne met Garrett James and his cousin Patrick, she has been tackling one particular piece of the prodigy puzzle: Does the cousin's autism have something to do with the prodigy's talent? The answer seems to be yes. Many of the prodigies have autistic relatives. Brothers. Sisters. Uncles. Grandmothers. Some have autism in every twig and branch of the family tree. The prodigies themselves—all of them—have autistic characteristics, such as extraordinary attention to detail and a tendency toward obsession. They draw on these traits to rocket to the top of their fields; these attributes are essential to their success. Prodigies and autists may even have a genetic link in common, a mutation on chromosome 1 that some prodigies and autists (but not their non-prodigious, non-autistic relatives) share.

This connection is fascinating; it offers an unexpected perspective on the riddle of the prodigies' talent and an intriguing take on what drives children to hone their skills with laser-like focus and intensity. But understanding child prodigies' abilities is only the first step; the next is to find out whether this connection could improve our under-standing of autism. Doing that requires investigating why it is that the prodigies have the strengths associated with autism but not the challenges.

The answer may be in the prodigies' genes. The prodigies and the autists appear to share genes on chromosome 1—a common foundation. But what if the two also have critical genetic differences? What if there is something about the prodigies' genes that protects them from autism's social and communication deficits but leaves the heightened attention to detail, astounding memory, and passionate interests of autism in place? If the prodigies and the autists share a genetic mutation *and* have important genetic differences, maybe studying the prodigies could unlock a piece of autism's infamously complex genetic architecture. It could mean that a breakthrough in autism research will come not from studying autists, but from studying child prodigies.

This sort of thinking has recently lit other areas of medical research on fire. For decades, scientists sought to learn more about conditions ranging from diabetes to heart disease by studying those who *have* diabetes or heart disease. Scientists rooted through the underlying biology of those with various medical conditions to try to understand what, exactly, was going wrong.

But recently scientists have begun looking in the other direction as well. Instead of focusing solely on those who are sick, they've taken a keen interest in those who are well—especially those who are at high risk for a particular disease, due to genes or lifestyle or both, but don't develop it. The idea is that if the scientists can isolate whatever it is that protects these inexplicably healthy individuals from the disease in question, perhaps they can use that knowledge to help those who actually have the disease.

HIV research is a good example of how this works in practice. To explore the potential implications of this strategy, it's worth taking a detour through the history of HIV research and considering whether a similar road map could lead autism research in an interesting direction.

—

Not long after AIDS was identified in the early 1980s, scientists began studying individuals with a high likelihood of contracting the virus: homosexual men with many sexual partners; prostitutes in Kenya and Gambia; adults with a history of intravenous drug use. As these studies progressed, the scientists found something unexpected. Even among these extremely high-risk groups, there were pockets of people in whom the virus never took hold.

Take Erich Fuchs and Stephen Crohn, for example. Erich had been exposed to HIV multiple times through unprotected sex with HIV-positive men. Stephen's partner, Jerry Green, was one of the first people in the United States to die of AIDS. Both men fully expected to contract HIV. They thought it was only a matter of time. *But it never happened.*

Mystified by their inexplicable luck, Erich and Stephen separately contacted research institutions and doctors, volunteering themselves as subjects. Eventually, both men crossed paths with Bill Paxton, then a postdoc at the Aaron Diamond AIDS Research Center in New York, who was seeking out a group of high-risk, uninfected individuals. "It was clear the minute I met Erich and Steve—these people should be HIV positive," Paxton recalled.

Paxton and his colleagues included Erich and Stephen in a group of twenty-five high-risk, HIV-negative individuals and homed in on their underlying biology. The idea was to extract their blood, infect their cells with HIV in the lab, and then figure out how they managed to shut down the virus's ability to replicate. But it was difficult to study virus replication in Erich and Stephen; their cells were nearly impossible to infect with HIV in the first place. The scientists used ever-increasing doses of HIV, but still the infection didn't take hold. "We repeated and repeated and repeated because we were throwing *very* high dosage of virus on these CD4+ cells, and we did not see infection," Paxton said. "We never thought for a minute that would be the outcome."

When the team later investigated the basis for this resistance, they stumbled onto a gem of a finding: Erich and Stephen shared a genetic mutation. Both had two copies (in scientific terms, they were homozygous) of what became known as CCR5-Delta32. This genetic "defect" prevented these individuals from producing a protein, CCR5, that usually sits on the outside of an individual's T cells and serves as a main entry point for the HIV virus. CCR5 isn't essential to every strain of HIV—some varieties rely upon other means to enter the cell—but for those strains of HIV that rely on CCR5, a double dose of the mutation makes patients' cells essentially impenetrable to HIV; the virus washes right through their bodies.

A search for this mutation in other HIV research subjects yielded some lopsided statistics. Across three studies, scientists found a few dozen HIV-negative individuals with two copies of this mutation. But no one infected with HIV—not a single one of the 2,741 HIV-positive individuals in these studies—had a double copy of CCR5-Delta32. It seemed that what Paxton and his colleagues had found in the lab was no fluke; the mutation was protecting its carriers from the virus.

Critically, as far as the scientists could tell, there was no price to be paid for this mutation. It resulted in high resistance to HIV, but it wasn't associated with any obvious defects or deficits. It was a genetic gift, no strings attached.

The treatment possibilities were explosive. In the aftermath of CCR5-Delta32's discovery, scientists set to work creating pharmaceuticals that could mimic its function. In 2007, the FDA approved maraviroc, the first of these drugs. It binds to CCR5 and blocks HIV from using it as a cell entryway.

But pharmaceuticals were only the beginning; identifying CCR5-Delta32 had even more astounding implications. Perhaps no one knows this better than Timothy Ray Brown, the first (and, so far, only) person ever cured of HIV—a status he would never have assumed if scientists hadn't taken the unusual step of studying a disease by zeroing in on those who don't have it.

⌐

Timothy was a twenty-nine-year-old American expat living in Europe when he was diagnosed with HIV in the mid-1990s.

At first, he was terrified; a former partner predicted he probably had only two years to live. But Timothy's timing was lucky. New pharmaceuticals that effectively managed HIV were coming onto the market around the same time that Timothy was diagnosed. He took the new drugs, his symptoms never really worsened, and he soon realized that HIV might not be a death sentence after all. Timothy continued building his life in Berlin. He got a job translating documents from German into English, and he hit the town at night. His health seemed almost normal.

In June 2006, when Timothy was forty, he flew to New York for a friend's wedding. He made it through the weekend's events—the party the night before, the wedding itself, the dim sum brunch the next day—but he felt exhausted the entire time. That Monday, after he returned to Berlin, he rode his bike ten miles to work, as he often did. The ride took much longer than usual, though. At lunchtime, Timothy tried to ride his bike to a restaurant half a mile away, but he made it only halfway before he was overcome by fatigue.

Timothy was soon diagnosed with acute myeloid leukemia, a rapidly progressing form of cancer. His oncologist contacted a Berlin hospital, where, by chance, he got Dr. Gero Hütter on the phone. Dr. Hütter said to send Timothy in, and he started him on chemotherapy.

In the meantime, Dr. Hütter initiated a search for a potential stem cell donor in case a transplant was necessary. A large number of potential matches turned up; there were eighty matches at the German Bone Marrow Donor Center alone.

Dr. Hütter, who was thirty-seven at the time, specialized in cancer, not HIV, but he remembered learning about CCR5-Delta32 in medical school. The surprisingly large number of matches led Dr. Hütter to wonder if they should perhaps be picky about the donor. Did any of Timothy's potential matches have two copies of

CCR5-Delta32? Could stem cells from such a donor cure Timothy's leukemia *and* his HIV?

Over the course of four months, Dr. Hütter's team tested potential donors for the CCR5-Delta32 mutation. On their sixty-first attempt, they found a match who was homozygous for CCR5-Delta32—just like Erich and Stephen. That individual agreed to donate if and when the time came. Timothy was already heterozygous for CCR5-Delta32 (meaning he had a single copy but not the magic-bullet double copy). It was a long shot, but the hope was that the transplant would leave him effectively homozygous for the mutation and potentially protected from HIV.

As far as Dr. Hütter knew, it would be a first-of-its-kind trial. "We have no experiences of this," Dr. Hütter recalled. "There were no cases published before, and there were also no animal results, and so we have totally no idea what will happen if we do this."

But Timothy was ambivalent about the procedure. After chemotherapy, his leukemia was in remission. Stem cell transplants are risky. His medication kept his HIV under control, and the idea of actually *curing* his HIV seemed far-fetched. Timothy initially refused the stem cell transplant. But when his leukemia returned at the end of 2006, he didn't see any way around it.

⌐

On February 6, 2007, eight months after he was diagnosed with leukemia, Timothy underwent surgery. He received the stem cells from the donor with two copies of the CCR5-Delta32 mutation. Just beforehand, he stopped taking his HIV medication. The operation went smoothly, and there were no serious complications.

Early on, a couple of tests detected HIV in Timothy's DNA. But soon all the tests were coming back negative; after a few months, there was no trace of HIV in Timothy's body, even though he hadn't taken his HIV medication since his surgery.

Five and a half months after the transplant, the doctors performed a colonoscopy. There was no sign of HIV in Timothy's rectal mucosa (the inner lining of the rectum), a potential hiding place for a viral reservoir. Twenty months after Timothy had stopped taking his medication, still no virus.

Timothy went back to work for the translation company. He rode his bike. He started working out again and could finally develop muscles because he no longer had wasting syndrome.

But his reprieve was short-lived. Timothy's leukemia came raging back at the end of the year. The doctors decided on another transplant—which would, again, be a risky procedure—from the same donor, the one with the double CCR5-Delta32 mutation.

As Timothy, who is chronically understated, put it, "That one didn't go so well." As he remembers, his blood platelet count plummeted. He began seeing black spots and temporarily lost his vision. During a conversation about a business venture, his words came out jumbled. The doctors suspected some sort of neurological disorder; they conducted an MRI and then biopsied Timothy's brain. For a while, Timothy lost the ability to walk.

Timothy began what would turn into more than a year of physical therapy. He eventually moved back to the United States, and his recovery is ongoing. His speech has returned to normal, though long conversations can still tire him. His condition improves daily, though he has residual balance problems and doesn't walk quite as he did before the procedure. But as of the spring of 2015, it had been eight years since he took any sort of medication to control his HIV, and he's still HIV negative.

⟵

For a long time, no one used the word "cured."

Dr. Hütter wrote up Timothy's case and submitted the paper to the *New England Journal of Medicine*. It was rejected. He applied to present Timothy's case at the 2008 Conference on Retroviruses and Opportunistic Infections in Boston, but the conference organizers allowed him only a poster on which to describe his results. It didn't create much of a stir.

A few weeks later, an AIDS researcher read about Dr. Hütter's work and invited him to present Timothy's case to a small group of scientists in September 2008, over a year and a half after Timothy's first stem cell transplant. Some of the audience thought that HIV had to be hiding somewhere in Timothy's body; they agreed, though, that he was "functionally cured."

The *New England Journal of Medicine* reconsidered Dr. Hütter's paper describing Timothy's case and published it in 2009. But Dr. Hütter still didn't use the word "cure." Instead, he described Timothy as having "long-term control of HIV."

Even Timothy, at first, avoided declaring himself free from HIV. "I was actually afraid of using the word 'cured' for a long time because I felt like it might give people false hope. At that point, I didn't really know for sure that I was cured, and I didn't want it to come back that I actually do have HIV," Timothy said.

As time went on and an ever-increasing number of tests failed to find HIV in Timothy, Dr. Hütter grew bolder. In a 2011 paper published three and a half years after Timothy stopped taking his HIV medication, Dr. Hütter and his colleagues declared it "reasonable to conclude that cure of HIV infection has been achieved in this patient." Two months later, the San Francisco AIDS Foundation held a forum with what would have been, just a few years before, an unthinkable title: "Is 'Cure' Still a Four-Letter Word?"

In June 2012, talk of a cure suffered a bit of a setback. At a conference in Spain, a researcher reported that his team had found traces of HIV in Timothy's body using ultrasensitive tests; two other teams of researchers reported similar results. However, these traces didn't match the HIV found in Timothy's body before the stem cell transplant, and some of these researchers cautioned that it might have been a false positive due to laboratory contamination. Either way, it's undisputed that in the eight years since Timothy's stem cell transplant, he's never taken his HIV medication, and there's no sign that the virus is replicating in his body.

Since Timothy's transplant, the same method has been tried in a few other cases. So far, no second cure. But Timothy is convinced that his case is proof that curing HIV is possible, and he has taken up the cause of advocating for finding a cure that could work for more people. He has vowed not to stop until HIV is cured.

⟶

Stem cell transplants aren't a feasible option for most people living with HIV. But Timothy Ray Brown's case was proof that a cure was possible, and scientists took note.

Inspired by his eradication of HIV, some researchers tried to leverage his stunning CCR5-Delta32 results in another direction: What if, instead of merely developing pharmaceuticals, they *edited the cells* of those with HIV to mimic the effect of the mutation?

In 2014, a group of researchers reported that they had tried. They had extracted blood from a group of HIV-positive patients and used zinc-finger nucleases (a sort of "molecular scissors") to sever the T cells' CCR5 genes. The idea was to dismantle both copies of the gene so that the patients' treated cells would be immune to attack by the HIV virus and then return those cells to the patients.

The modified cells could be detected in every patient. The modification, moreover, seemed to have worked: when the study participants stopped taking their HIV medication, the treated cells fared better than the nontreated cells.

The scientists found something else, too: Patient 205.

After Patient 205 stopped taking his HIV medication, it took six weeks for his viral load to increase. But then something unexpected happened. Without medication or any further intervention, his viral load began to decline. By the time the treatment intervention was over, a point at which the research protocol called for all participants to resume taking their HIV medication, his viral load was undetectable.

Intriguingly, Patient 205 was already heterozygous for CCR5-Delta32, just like Timothy Ray Brown before his stem cell transplant. Patient 205's built-in single mutation meant that he effectively received a bigger dose of treated cells; while the molecular scissors the team used to alter the patients' cells might have disabled one or both copies of the relevant gene in each of the other participants' cells, every time the scientists disabled one of Patient 205's CCR5 genes, he already had an inborn mutated copy to match, giving him a double dose of the mutation.

HIV isn't an isolated case study of the power of studying those *without* a particular disease or disorder. By using this somewhat counterintuitive method, scientists have fine-tuned our understanding of other diseases and have even identified other beneficial mutations—the new holy grail of medical research. In one recent study, for example, scientists investigating type 2 diabetes compared the genes of those who had type 2 diabetes with the genes of those who were old and overweight but still (inexplicably) healthy. This led them to the SLC30A8 gene, which is related to insulin production. It turns out that those with a mutation that inactivates one copy of this gene have a 65 percent reduction in risk for developing type 2 diabetes. Scientists studying heart disease sequenced the PCSK9 gene in individuals with extremely low LDL cholesterol (the bad kind) and identified mutations tied to a reduced risk of coronary heart disease. The identification of beneficial mutations that reduce the risk of heart disease has the pharmaceutical industry salivating. So promising is this line of research that a team of scientists recently launched the Resilience Project, a program dedicated to identifying more such beneficial mutations.

These beneficial mutations would never have been discovered, and the associated treatments likely wouldn't have been developed, if scientists hadn't scoured the DNA of people who were well, despite being at high risk for various diseases. The at-risk but inexplicably healthy subjects of the CCR5-Delta32 studies are the prodigies of the HIV world; researchers studied them in an effort to help their "cousins"—those who contracted the virus.

⟵

Now, let's not get too far ahead of ourselves. These efforts to better understand those who are well have produced some spectacular results—but autism is not HIV, type 2 diabetes, or heart disease. There are a couple of important reasons why autism is different.

First, not everyone agrees that you ought to think about "curing" autism in the same way you think about curing diabetes or heart disease; there's real debate about how best to support autists and their families. Some advocates argue that we should view conditions like autism as neurological *variations,* not neurological disorders. Autism then is a distinct combination of strengths and weaknesses and a part of the individual's personhood. As Jim Sinclair, one of the founders of Autism Network International, an autism advocacy organization run by the autistic, put it in his 1993 essay, "Don't Mourn for Us,"

> Autism isn't something a person *has,* or a "shell" that a person is trapped inside. There's no normal child hidden behind the autism. Autism is a way of being. It is *pervasive;* it colors every experience, every sensation, perception, thought, emotion, and encounter, every aspect of existence. It is not possible to separate the autism from the person—and if it were possible, the person you'd have left would not be the same person you started with.

From this perspective, focusing on the search for autism's genetic roots could be misguided (as could efforts to develop pharmaceutical treatments, which are often tied to this sort of work). Autists may not view their autism negatively. Autists also can and do make great contributions to society, and as Jacob Barnett observed, they may be able to do so not in spite of their autism but *because* of it. And every dollar spent on analyzing genes is a dollar not spent on accommodations, support, and efforts to increase sensitivity that could help autists now. As Julia Bascom, deputy executive director of the Autistic Self Advocacy Network, put it, "The biggest barrier the autistic community faces is not our autism, but a society which is ignorant, unaccommodating, and often actively hostile to people who are different, people with disabilities, and autistic people."

Others are equally adamant about the necessity and urgency of finding effective ways to treat autism. It's a disorder, they believe, and parents should do all they can to help their children fight against it. Those who argue for acceptance over intervention, they often claim, are "high functioning": they don't appreciate the difficulties faced by those with more severe autism.

The second challenge with using this approach in autism research involves both the methodological difficulties and the heterogeneity of autism genetics. Child prodigies are rare, so scientists attempting to study their genes are stuck with a small sample size. Similar work *has* been done in genetics studies involving siblings of autists (the autists' genes are compared with those of their non-autistic siblings), and while this approach has helped identify autism-linked genes, it hasn't yielded any genetic variants that seem beneficial. Autism has such complicated, knotty underlying genetics that it may not be possible to tease out anything useful. "So far, none of these single-gene studies has given us anything that's in some sense actionable to say *okay,* we can take this, block that gene, or further goose up the effect of that gene, and it's gonna get us somewhere," Bruce Cuthbert, the acting director of the National Institute of Mental Health and the director of the Research Domain Criteria project (which we'll get to shortly), said. "That just hasn't proven to be the case."

But that doesn't mean it's not worth exploring. The debate over the nature of autism is important and should be approached with great sensitivity, but you don't need to believe that autism should be "cured" to appreciate the power of studying those who are well from a research perspective. The scientists who discovered that certain SLC30A8 mutations may lower type 2 diabetes risk, for example, challenged the conventional wisdom about that gene, which had previously been associated with an *increased* diabetes risk. The first time the team submitted their findings for publication, the paper was rejected. "It was so at odds with the previous knowledge of how this gene had worked," Jason Flannick, the lead author of the SLC30A8 study, said. "This is a totally new hypothesis that has very strong genetic data behind it, but it's definitely not the end of the story; it's the start of what will be a very long period of work." This sort of insight seems like something that could be quite helpful to autists, autists' families, and scientists. As the prominent autism researcher Geraldine Dawson put it in the general context of autism genetics studies:

> We're not really trying to cure autism in the sense that we think autism is something that you absolutely want to get rid of, because autism actually comes with gifts and unique differences that I think are really special and very important to have as part of our human society. Really, what we want to be able to do is to help each individual with autism reach their full potential—to be able to communicate and to be able to use the unique talents and gifts that they have and also not to suffer from some of the medical comorbidities that go along with autism.

To this end, the prodigies have something important in common with Erich Fuchs and Stephen Crohn. Given their family histories, the prodigies could be considered at high risk for autism, just as Erich and Stephen were at high risk for HIV, *and* unlike the typically developing siblings of autists, the prodigies all demonstrate some truly extreme autism-linked behaviors and cognitive abilities. But just like Erich and Stephen, who never contracted HIV, the prodigies don't have the deficits associated with autism.

From this perspective, maybe the prodigies aren't just a marvelous curiosity. Maybe they're a potential Rosetta stone for some variations of autism. And as for autism's complexity and heterogeneity, there's at least one prominent organization where those issues are high on the research agenda.

⟶

Bruce Cuthbert has gray hair, blue eyes, and an oval face. When he smiles, which he does often, he resembles a midwestern news broadcaster.

His office is on the eighth floor of the Neuroscience Center in Rockville, Maryland, a building that houses the National Institute of Mental Health's headquarters. The space is nondescript: cream walls, gray overhead bins, books lining the top of the shelves. It hardly looks like the staging ground for a mental health revolution. But it's from this office that Cuthbert is leading the Research Domain Criteria project, an effort more commonly known as RDoC—NIMH's answer to the mismatch between symptoms-based *DSM-5* diagnoses and the underlying biological reality.

This effort leaped into the public spotlight in 2013 when Thomas Insel, then the director of NIMH, slammed the new edition of the *DSM* as little more than a dictionary. It described groups of symptoms, he said, but didn't actually *diagnose* anything; the symptom clusters weren't rooted in the underlying biology. Using symptoms to diagnose mental illnesses, he argued, was like diagnosing diseases of the body based on the type of chest pain or severity of fever. "As long as the research community takes the D.S.M. to be a bible, we'll never make progress," Insel told the *New York Times*.

It was a sentiment that had been brewing for years. From Cuthbert's perspective, there had always been something a bit strange about the *DSM* categories. When the third edition, a revision aimed at establishing more reliable diagnoses for a field that was struggling with consistency, came out in 1980, Cuthbert, who was then a psychology researcher with the U.S. Army Medical Services Corps, recalls being flummoxed. "It was all like a magical mystery tour to us," Cuthbert recalled. "What is all this stuff? Where did they get all this stuff? To me, they never really did make all that much sense from a natural science point of view."

But when Cuthbert began his first stint at NIMH in 1998 (he was the chief of the Adult Psychopathology and Prevention Research Branch from 1999 to 2005), he realized that many researchers didn't share his skepticism about the *DSM* categories. These scientists treated them as hard-and-fast diagnoses that described distinct disorders.

Strict adherence to *DSM* diagnoses was problematic for research; scientists grouped their subjects according to these categories despite growing evidence that the clusters of symptoms didn't map onto any single underlying disorder.

It was even more problematic for attempts to identify treatments, the effectiveness of which varied widely among individuals who all theoretically had the same disorder. Pharmaceutical development in particular suffered. Drug development can proceed only if scientists know what they're trying to target; with *DSM* disorders' underlying biology murky, pursuing pharmaceuticals was a fool's errand. A number of companies pulled back from developing psychiatric drugs.

NIMH scientists eventually set out to tackle the problem. In NIMH's 2008 strategic plan, the organization declared its intent to develop a new classification system for brain disorders that took into account behaviors *and* biology; that's RDoC.

RDoC casts aside current diagnoses like autism and schizophrenia. Instead of using these recognized—but, for research purposes, confusing—terms, RDoC breaks brain functioning down into broad constructs like "negative valence systems," which includes fear and anxiety, and "cognitive systems," which includes attention and language. These categories cut across current diagnoses in an effort to get at the underlying mechanisms that result in brain disorders and, from there, behavioral abnormalities.

The idea is that by untangling the roots of brain disorders, scientists can develop personalized, targeted treatments. Perhaps with a better understanding of the conditions we have long known as autism or schizophrenia, depression or obsessive-compulsive disorder, scientists can open up the door to improved behavioral, pharmaceutical, and even genetic treatments.

"We have treatments, but they're not nearly as precise as we want," Cuthbert said. "So if you really want to do a better job of diagnosing and treating people, it's clear that we are going to have to face up to the heterogeneity that exists with all of our disorders and move in this precision medicine direction, and that's really what this Research Domain Criteria thing is all about."

At this point, the minds behind RDoC haven't explicitly set out to emphasize the study of those at high risk for a particular condition but not actually affected by it. But they are attempting to broaden the range of behaviors researchers study beyond those that would qualify for a formal diagnosis. The idea is to acknowledge that many traits, like anxiety, exist on a continuum. There is no clean on-off switch demarcating the point that distinguishes those who have an anxiety disorder from those who do not. But because scientists often study only subjects who qualify for a diagnosis, there is little research on those whose symptoms put them on, but not quite over, the diagnostic threshold.

It's a start, but science still has a long way to go before we can unravel terms like "autism" and "schizophrenia" to see what really lies beneath. In the meantime, the possibility of approaching autism from a novel angle beckons. After all, scientists have already made some headway toward identifying a mutation that seems to result in lower levels of anxiety and enhanced fear extinction (the ability to forget a learned fear response). Child prodigies, a group of individuals who seem to be at high risk for autism but who don't have the typical social and communication difficulties, seem like another particularly promising starting point.

The prodigy genetics research is ongoing. The Ohio State team still needs to pin down the chromosome 1 mutation that prodigies and autists seem to share. Guy Rouleau and the Canadian team are hunting for a de novo mutation that contributes to prodigious talent. It's possible that neither team will find anything of interest to autism researchers, but it's also possible that they will.

It's only by actually studying child prodigies, a group long relegated to the research sidelines, that we'll find out. If the connection with autism bears out, if prodigies really can point the way toward an improved understanding of autism, maybe child prodigies aren't so much a mystery anymore. Maybe they're the beginning of an answer.

Epilogue

A Wide-Open Future

What happens when child prodigies grow up?

There are plenty of popular reports—and conjectures—about the fate of child prodigies. Some assume that, like Mozart, the prodigies' spectacular early achievements destine them for the pinnacle of their professions. Others buy into the "early ripe, early rot" perspective. The idea there is that the prodigies, besieged by pushy parents and excessively exposed to the media, lacking friends their own age or the grounding to weather the natural ups and downs of a long career, will break down or fade into obscurity.

The real answer is that no one knows. Just as autism in adulthood is relatively under-studied, so, too, are child prodigies' grown-up years.

What little can be gleaned from the academic literature suggests that there's a wide range of outcomes for prodigies (just as there is for everyone else), and that most fall between the extremes of global stardom and the sort of failure predicted by early ripe, early rot. A small study of eight chess prodigies, for example, found that most became eminent chess players and one became a world champion. David Feldman has lost touch with two of the six prodigies he studied closely. The other four, though, have mostly carved out careers within their original fields of interest. The music prodigy graduated from Juilliard and is a solo violinist. The writing prodigy graduated from Harvard and writes about music professionally. One of the chess prodigies has become an attorney, and the jack-of-all-trades prodigy is a musician and a freelance computer consultant. There's no one in his group of six who has achieved a household-name level of acclaim, but in fields as competitive as music and writing (or as relatively obscure as chess), what are the chances that there would be?

Child prodigies certainly have some advantages as they transition into adulthood. They have extreme work ethics and the confidence of having already developed a noteworthy ability. They have often found a measure of fame that may open up other opportunities.

Children who have already achieved much also face some unique challenges, though. The media attention may dwindle, confusing adolescents who have become accustomed to constant praise. Pursuits that once came naturally may prove more difficult, or the prodigies may find that their interests no longer align with conventional pictures of success. Critics can seem almost gleeful when a prodigy falls off his or her childhood pedestal and into something closer to normality.

Most of the prodigies described in this book are still quite young. Some are still pursuing those passions that first seized them during childhood; others have moved on to other interests. Some will have outstanding adult careers that seem to fulfill the promise of their early abilities. Some may try for success and fail. Others will opt for quieter lives, different paths, and their own unique adventures.

But what the child prodigies make of their talents is only one aspect of their lives. After all, child prodigies aren't just highly accomplished individuals. They're funny and charming and driven, and they have incredibly generous hearts. It's been an amazing privilege to get to know these kids, spend time with their families, and offer a glimpse into their world.

For now, the best anyone can say with certainty is that these children are neither doomed to fail nor predestined for adult success. Here's an update on their lives since we left off with their stories.

Garrett James

The tiny Louisiana strummer is still a wholehearted musician. He has released several CDs and tours frequently.

Greg Grossman

The Greg Grossman cooking cyclone has never slowed.

When Greg was twenty, he became the executive chef of Beautique, a new, high-end Southampton restaurant and the sister location to a popular Manhattan eatery. On the restaurant's busiest nights, his team served four hundred to five hundred people.

He's a partner and co-founder of Kettlebell Kitchen, an organization that delivers Paleo meals to more than two hundred gyms across New York and New Jersey. It has expanded rapidly and delivers more than fifteen thousand meals per week.

Greg and his partners have two new restaurants in development in Manhattan—a small plates Mediterranean restaurant planned for Midtown and an East Village microbrewery. Both are scheduled to open shortly.

He still loves exploring ingredients, cooking products, and new techniques, and he of course loves eating. He can't get enough of the challenge of running a kitchen and the close camaraderie that emerges with co-workers engaged in a true team effort.

"It's a different battle every day. It's constantly changing, and you have to stay on your feet," Greg said. "It's very exciting, and it never really gets old."

Jonathan Russell

Jonathan is studying music composition at NYU. Since he began college, he's released a CD of original compositions and has scored several new projects, including a feature film and a Web TV show for kids. He still performs in Central Park.

"I have more fun composing than performing," Jonathan said. "When I compose, I can kind of just get into my own head and enjoy myself without having to worry about anything, and I certainly have stamina to do that a lot longer. I'll sit down, and I'll write for six hours at a time. I can't perform for six hours, nonstop—it's just too much."

Lauren Voiers

Lauren thrived as a professional artist for the first couple of years after she finished high school. She jaunted around the country doing group and solo exhibitions, made the media rounds, and completed multiple commissioned works. One of her paintings, *Peace & Harmony,* was made into a sculpture and installed in Liverpool in honor of what would have been John Lennon's seventieth birthday.

She put her art career on pause when she parted ways with her agent. For several years, Lauren stopped painting.

She has since enrolled at Santa Monica College, where she's studying fine art. She's resumed painting and often creates dramatic, bold-colored pieces; she's experimenting with pen and pencil drawings and has a burgeoning interest in photography.

Richard Wawro

Richard's legacy lives on through his artwork and through his family. One of Mike's daughters recently redid Richard's Web site as a birthday gift to her father. Mike still hears from people through the Web site who own Richard's artwork and talk about how important and inspiring it is to them, and he receives messages from people who are deeply moved by Richard's story.

Jacob Barnett

Jacob reveled in every aspect of college. He loved his classes. He liked getting to know the other students and even tutored some of them: the only prerequisite was that they bring spoons to partake in the giant tubs of peanut butter he brought along to snack on during study sessions.

After his freshman year, he worked as a paid research assistant in quantum physics at IUPUI as part of an undergraduate program; during this time, he tackled a previously unsolved math problem. Afterward, he and his mentor coauthored a paper that was published in a noted, peer-reviewed physics journal. It's titled "Origin of Maximal Symmetry Breaking in Even P T-Symmetric Lattices."

At fifteen, he enrolled at the Perimeter Institute for Theoretical Physics in Ontario. The Barnetts sold their home in Indiana and moved to Canada, and Jacob is now a Ph.D. candidate. His TEDx-Teen talk, "Forget What You Know," in which he urges listeners to stop learning and start thinking and creating, has been viewed more than six million times.

"I think he is just happy being him. I don't know if in the future we're gonna make all of the decisions that everybody expects us to make," his mother, Kristine, said. "I think that the compass that's always gonna drive Jacob is just, is he doing what he loves to do? Then we'll be doing that thing."

Jourdan Urbach

Jourdan's last two years at Yale had the same frenzied pace that marked his years growing up on Long Island. He was selected to score a short film at the Columbia University Film Festival. He built and ran a recording studio, performed as part of a Haiti benefit concert, and served as a United Nations Art for Peace goodwill ambassador. He found out that the paper he had coauthored with Joanne on child prodigies would be published while he was at the shooting range with the Yale Pistol & Rifle Club.

At twenty, he received a Jefferson Award for Outstanding Service for his philanthropic work and a Tribeca Disruptive Innovation Award alongside Jack Dorsey, Justin Bieber, and others.

He graduated from Yale a year early (in part through credit for high school AP classes; in part by stuffing a large number of courses into a single semester) and began a one-year stint as the national director of the Jefferson Awards.

Jourdan still plays the violin, but these days it's just for fun (or fund-raising).

Ocho, the eight-second social video platform that he co-founded, received $1.65 million in seed funding by the end of 2014 from a group of investors that included Mark Cuban, the billionaire investor of *Shark Tank* fame.

"I wanted to build products that had a mandate to exist, and I think Ocho has a mandate to exist. There needs to be a way to share your life with people through videos, not just through photos," Jourdan said. "I won't work on products where I need to win on marketing. That's my nonprofit background coming into it. I feel it's immoral. I think it's immoral to win on marketing in the nonprofit world, and I think that it is wrong and probably dumb to work for a company that wins based on marketing, even if it's a for-profit."

Josh and Zac Tiessen

Josh and Zac Tiessen are both blossoming professionally.

Josh was named one of Canada's Top 20 Under 20 for his art and charity work. By the time he was nineteen, his gallery included paintings priced over C$17,000. He recently sold a painting for over C$23,000. He exhibits his work frequently, and his painting *Ahoy Sleeper,* a haunting depiction of a diver emerging from the water at night, won the Creative Achievement Award at a selective International Guild of Realism exhibition in Charleston. As his mother, Julie, noted, "His only challenge as a teenager at these events has been booking into hotels, renting cars, and being underage at his own wine and cheese exhibition openings!"

Zac has composed more than thirty original songs; he put four of those songs on his first solo CD, which he released when he was seventeen. His YouTube channel has been steadily growing, and he's received a number of endorsements from music equipment companies. A few months after his CD release, he opened a concert for Animals as Leaders in Toronto with "a blistering three song set."

He completed a specialist certificate in guitar skills through the Berklee College of Music online program. Zac had hoped to apply for the full-time program at Berklee, but his concussion specialist nixed that idea. Instead, Zac bought the textbooks used in the Berklee bachelor's program and worked his way through them independently. He's now studying master's level jazz theory. At eighteen, he was written up in *Guitar World* for his work on the eight-string guitar; the magazine also recently spotlighted one of his playthroughs.

Personally, the road has been much more difficult for the Tiessens. The boys' Lyme symptoms worsened over time. Josh noticed some mental fogginess, flu-like fatigue, and physical discomfort. Zac, too, experienced some difficulty concentrating along with fatigue, mood swings, and insomnia, though in his case it was less clear whether his symptoms stemmed from Lyme or his concussions.

The entire Tiessen family was treated for chronic Lyme disease at the Sponaugle Wellness Institute in Florida. The doctors there also discovered signs of exposure to mold and industrial toxins in their blood. Their friends and family rallied around their cause and raised more than $250,000 to help pay for the family's care. Even in the midst of the temporary relocation to Florida and ongoing medical treatment, Zac stayed up all night to finish a music video and opened for a progressive band in Orlando; Josh continued to paint several hours a day.

Autumn de Forest

Autumn travels frequently these days—New Orleans, Orlando, Scottsdale, Minneapolis, San Antonio, Boston—discussing her work with art collectors. She recently visited Savoy Elementary School in Washington, D.C., in connection with Turnaround Arts, an organization that provides arts education programs to low-performing schools.

She's still a media darling, was recently featured on a Times Square billboard in New York as part of P.S. from Aéropostale's "epic kids" campaign, and donated a painting for a world hunger charity fundraiser.

And she still loves the way that painting lets her express herself and her imagination. "I've never taken lessons; it's all come from me," Autumn said during a Disney Citizen Kid spot. "I just go for it. I let my heart go."

Ping Lian Yeak

Ping Lian continues to paint in Sydney. Rosa C. Martinez, the founder of Strokes of Genius, has helped him to secure representation at a Manhattan art gallery. His mother is working on a book about raising Ping Lian.

Alex and William

Lucie's children continue to thrive, and they continue to do so outside the spotlight.

Her older son, Alex, has completed fifth grade. He's a strong athlete, loves skiing, and recently took up gymnastics. He also enjoys music, especially (to Lucie's dismay) classic rock and heavy metal.

He's voraciously interested in the mechanics of how things work, and he's a Lego fanatic. He builds quickly, and he does it every day, often in the mornings before Lucie wakes up. He has exhibited some of his builds at Lego conventions.

All signs of his autism (other than his keen eye for detail) have disappeared. He's a happy-go-lucky kid. His latest report card, including the behavior report, was excellent. He's a popular student, participates in class discussions, and was chosen to speak at a school assembly based on his communication skills.

Alex's little brother, William, technically just completed fourth grade. But he's a couple of years ahead of grade level in most subjects, and he's almost finished the tenth-grade math curriculum. In a math competition for eighth-grade students, he placed first in the school.

His mind continues to amaze—and, often, delight—his family, friends, and teachers. When William was seven and taking a seventh-grade math class, another student asked whether it was theoretically possible to raise an exponent by another exponent—2 squared, to the power of 3, to the power of 4, for example. His teacher, Josh, answered that it would be a big number, but it was possible. A few minutes later, William began rattling off numbers: 16,777,216. Josh asked another student to check it on the calculator; that student verified the figure. William had correctly calculated the eight-digit answer in his head.

He computes ages in binary and hexadecimal. Once, when Lucie came downstairs, he told her the new population density (in person per square kilometer) of the upstairs of the house. During a conversation between Alex and Lucie about women's suffrage, William chimed in by reciting the year in which each of the Canadian provinces gave women the right to vote. The books on his bedside table include a treatise on infinity and an overview of physics concepts. He still plays the piano and has found some music apps that allow him to compose multi-instrument pieces. He's recently taken up writing computer code.

There are still challenges for William. Transitions between activities can be difficult; he requires prompting throughout the day.

But it's been a landmark year for him socially. He's developed several close friends who often request playdates. His anxiety level has dropped, and he's increasingly willing to try new things (Lucie whipped out her credit card the second he showed interest in jujitsu).

He, too, was singled out to speak at a school assembly based on his communication skills.

Alex and William both recently opted for a summer of personal projects and unstructured fun instead of day camps. The family has a pool and a trampoline, but Lucie has to prod them to go out and play; they still find the mental gymnastics they can do inside to be the most captivating sort of activity.

"Both kids are very bright, but my priority is that they're happy," Lucie said. "I don't care if they don't win a Nobel Prize. I just want them to be happy and to have a social circle that's supportive. I want them to be around people who are healthy for them and truly appreciate them for who they are. That's what I want in life for them; *that's* successful."

Acknowledgments

Writing this book introduced many wonderful people into our lives, and *The Prodigy's Cousin* would never have existed without them. Our lovely agent, Rachel Vogel, is a staunch advocate and trusted friend who offered a steady hand at every turn. We are grateful to everyone at Current who devoted so much time and thoughtfulness to shaping this book, including publisher Adrian Zackheim; executive editor Eric Nelson; our editors, Maria Gagliano, Emily Angell, and Jesse Mae-shiro; and editorial assistant Leah Trouwborst. Associate publisher and marketing director Will Weisser, publicity director Tara Gilbride, publicist Taylor Fleming, and publicity assistant Kaitlyn Boudah brought tremendous energy, creativity, and enthusiasm to this project.

Thank you to the other members of the Penguin Random House team who did so much for this book, including Karl Spurzem for the jacket design, Leonard Telesca for the interior design, Ingrid Sterner for her eagle eye during the copyedit, and Bonnie Soodek for handling subsidiary rights. Thank you to the managing editorial and production teams who worked hard to keep this book moving along, including senior production editor Bruce Giffords, executive managing editor Tricia Conley, senior production editor Jeannette Williams, and production manager Madeline Rohlin. We are also grateful to Jane Cavolina for pitching in with the endnotes.

Many friends and colleagues offered feedback at various stages of this project, including Soren Aandahl, Jennifer Caughey, Bryan Choi, Geraldine Cremin, Caitríona Palmer, Sonali Shah Pier, J. Maarten Troost, Diane Young, and Alexandra Zapruder. The members of the

East Side Writers in Providence and the D.C. Science Writers Association freelancers' group offered thoughtful comments and friendship. Joe Camoriano was an enthusiastic supporter of this project and created a knockout book trailer. To everyone at the Writers Room DC: thanks for making it fun.

Joanne's research couldn't have been done without the psychology department at Ohio State University; her research partners, including Chris Bartlett, Stephen Petrill, and Guy Rouleau; and funding from the Marci and Bill Ingram Research Fund for Autism Spectrum Disorders and Ohio State. We also want to acknowledge all the prodigies with whom Joanne has worked over the years. Only a few are profiled in this book, but each one is not only marvelously talented, but also charming, bighearted, and fun to hang out with.

We would like to thank the scientists, experts, journalists, and other professionals who shared their time and their insights with us, including Tania Barkat, Julia Bascom, Ryan Bogdan, Carrie Borrero, Bruce Cuthbert, Geraldine Dawson, Roger Detels, Doug Detterman, Deborah Fein, Jason Flannick, Uta Frith, Nancy Greenspan, Matthew Hill, Gero Hütter, Vicki Jenkins, Maria Kozhevnikov, Francis S. Lee, Jean Mercer, Laurent Mottron, Daniel Notterman, Adam Piore, Allan Snyder, Pablo Tebas, and Darold Treffert. For additional great stories, thank you to Audrey Curran, Gerald Mastellone, Michael Mastellone, Sara Mastellone, Donna McPeek, Gerri Ruthsatz, and Judy Ruthsatz.

We owe a special thank you to David Feldman, a smart, funny, gracious, and thoughtful human being who gave us an insider's view of his early days as a prodigy researcher and made every interview more fun than work. His thinking very much advanced our own, and his zest for his work inspired us at every turn. Our sincere appreciation to David for his boundless encouragement.

We're grateful to Jim for cheering us on at every step of this journey and for accompanying Joanne on that first fateful trip to Louisiana and her many adventures since. Thank you to Dan for poring over every draft, going to prodigious lengths to create more time in our lives for this book, and for the constant support. Kyle braved the icy roads on several harrowing prodigy road trips, and Ryan kept Joanne's computer running despite its best efforts. Katherine and Jack provided very creative edits to the manuscript and lots of squishy hugs. Our heartfelt gratitude to Bob and Dianne Stephens, Julian (Jim) Ruthsatz, and the late Marie Mastellone. Our whole family shared our enthusiasm for this story, making every part of the process all the more exciting.

Finally, a very warm and heartfelt thank you to those we profiled in this book, along with their families, teachers, and friends, who added great depth and perspective to their stories, including Alex, William, Lucie, Kathy, and Josh; Greg and Terre Grossman; Lauren, Doug, and Nancy Voiers; Jonathan Russell and Eve Weiss; the late Richard Wawro, Mike Wawro, and Laurence A. Becker; Jacob Barnett, Kristine Barnett, and Becky Pearson; Jourdan Urbach and Eric Drier; Josh Tiessen, Zac Tiessen, Julie Tiessen, and Valerie Jones; Autumn and Doug de Forest; Ping Lian Yeak, Sarah S. H. Lee, and Rosa C. Martinez; and Timothy Ray Brown and Dave Purdy. This book would not have been possible without their tireless cooperation, and we are deeply indebted to them for their kindness, generosity, and patience with our endless follow-up questions. Most of all, we are grateful for the way they welcomed us into their lives and shared their stories. We find them inspiring; we hope you do, too.

Notes

Epigraph

vii **"To call specific":** Ogden R. Lindsley, "Can Deficiency Produce Specific Superiority—the Challenge of the Idiot Savant," *Exceptional Children* 31, no. 5 (1965): 225–32.

Introduction

2 **Others had argued:** A few classic works and review articles on the importance of general intelligence, practice time, and skills specific to a particular field include Howard Gardner, *Frames of Mind: The Theory of Multiple Intelligences* (New York: Basic Books, 1983) (skills specific to a particular field); K. Anders Ericsson, Ralf Th. Krampe, and Clemens Tesch-Römer, "The Role of Deliberate Practice in the Acquisition of Expert Performance," *Psychological Review* 100, no. 3 (1993): 363–406 (practice time); Michael J. A. Howe, Jane W. Davidson, and John A. Sloboda, "Innate Talents: Reality or Myth?," *Behavioral and Brain Sciences* 21, no. 3 (1998): 339–442 (practice time); Frank L. Schmidt and John Hunter, "General Mental Ability in the World of Work: Occupational Attainment and Job Performance," *Journal of Personality and Social Psychology* 86, no. 1 (2004): 162–73 (general intelligence); Arthur R. Jensen, *The g Factor: The Science of Mental Ability* (Westport, Conn.: Praeger, 1998) (general intelligence).

2 **it was the combination:** For more information on this theory, see Douglas K. Detterman and Joanne Ruthsatz, "Toward a More Comprehensive Theory of Exceptional Abilities," *Journal for the Education of the Gifted* 22, no. 2 (1999): 148–58; Douglas K. Detterman and Joanne M. Ruthsatz, "The Importance of Individual Differences for Exceptional Achievement," in *Talent Development IV: Proceedings from the 1998 Henry B. and Jocelyn Wallace National Research Symposium on Talent Development,* ed. Nicholas Colangelo and Susan G. Assouline (Scottsdale, Ariz.: Great Potential Press, 2001), 135–54; Joanne Ruthsatz et al., "Becoming an Expert in the

Musical Domain: It Takes More Than Just Practice," *Intelligence* 36, no. 4 (2008): 330–38.

2 **College-level musicians:** Ruthsatz et al., "Becoming an Expert in the Musical Domain." This trend held when just the high school band members were considered: the higher the students' music achievement, the higher their IQ, domain-specific skills, and practice time. The same was true at the college level, although for these students only the relationship between practice time and achievement was statistically significant.

7 **Following the lead:** Thomas R. Insel and Bruce N. Cuthbert, "Brain Disorders? Precisely," *Science,* May 1, 2015.

Chapter 1: A Warehouse of a Mind

9 **She used to encounter it a lot:** The events in this chapter described by Lucie come from telephone interviews conducted on Sept. 4 and 12 and Oct. 12, 2014, July 12, 2015, and Oct. 20, 2015 (with occasional input from Mike); and e-mail. In addition, Lucie provided videos, photographs, and a written summary of Alex's and William's development dating from 2012.

10 **He consumed everything:** The "warehouse" description of William's mind comes from a multi-session psycho-educational assessment, Dec. 5, 2012, Dec. 13, 2012, and Jan. 22, 2013.

10 **Within months, Lucie sensed:** The details of Alex's development come from Lucie, as well as educational and medical reports, including his autism diagnosis, Nov. 22, 2005; his assessment for a preschool autism program, Feb. 23 and March 23, 2006; a psychological assessment, Jan. 22, 2009; a psycho-educational assessment, Jan. 6 and 20, 2010; and a psycho-educational assessment, March 19–21, 2013.

12 **Two months later, Alex was diagnosed with autism:** Autism diagnosis, Nov. 22, 2005.

19 **His diagnosis was stripped away:** Psychological assessment, Jan. 22, 2009.

20 **William wasn't as serious:** The details of William's development come from Lucie, as well as a telephone interview with Kathy (William's piano teacher) conducted on Oct. 10, 2014; a telephone interview with Josh (William's math teacher) conducted on Oct. 14, 2014; educational and medical reports, including his autism diagnosis, Nov. 21, 2007; an assessment for eligibility to participate in autism intervention, April 21, 2008; a psycho-educational assessment, March 6, April 3 and 17, 2010; a medical assessment, July 29, 2011; and a psycho-educational assessment, Dec. 5, 2012, Dec. 13, 2012, and Jan. 22, 2013.

22 **The doctor diagnosed William:** Autism diagnosis, Nov. 21, 2007.

22 **He echoed the words:** The information about William's echolalia is from his 2008 eligibility assessment for an autism intervention program. Lucie

remembers that William's speech was delayed and that even when he did begin talking, he didn't say much (he "certainly didn't have the gift of gab," as Lucie put it). She doesn't recall his echolalia, but at the time she was still putting a lot of blood, sweat, and tears into helping Alex, so she says it's possible that some things fell off the radar.

24 **On a standardized test:** Psycho-educational assessment, March 6, April 3 and 17, 2010.

25 **He began taking:** Medical assessment, July 29, 2011.

26 **"Yes," William piped up:** Lucie, unlike William, doesn't have perfect recall for atlas page numbers. She's not 100 percent certain that the actual pages he recited were 34 and 35.

Chapter 2: What *Is* a Prodigy?

32 **As articulated by Feldman:** David Feldman, "The Mysterious Case of Extreme Giftedness," in *The Gifted and the Talented: Their Education and Development,* ed. Harry Passow (Chicago: University of Chicago Press, 1979), 335–51, reprinted by the Davidson Institute for Talent Development. In this article, the standard for prodigiousness is "each child performs in his chosen field at the level of an adult professional before the age of ten." The "demanding field" aspect of the definition followed soon after. See David Feldman and Lynn T. Goldsmith, *Nature's Gambit* (New York: Basic Books, 1986).

32 **When he was a young assistant professor:** For an overview of Feldman's thinking regarding Piaget, see David Feldman, *Beyond Universals in Cognitive Development* (Norwood, N.J.: Ablex, 1980); and Feldman, "Mysterious Case of Extreme Giftedness."

33 **"So, I thought, okay":** David Feldman, interview, Feb. 21, 2014.

33 **"I know it when I see it":** This language was famously used in associate justice Potter Stewart's concurring opinion in *Jacobellis v. Ohio,* 378 U.S. 184 (1964), in reference to the difficulty of determining what constitutes hard-core pornography.

33 **In one of the earliest:** There was an earlier case study of a prodigy, but it has never been translated into English. See Alfred Binet, "La psychologie artistique de Tade Styka," *L'année psychologique* 15 (1908): 316–56. N. S. Leites, a Russian researcher, studied child prodigies as well, but these papers are also unavailable in English. Leites briefly references this work in an English-language article, "The Relationship Between the Developmental and the Individual in a Schoolchild's Aptitudes," *Soviet Psychology* 24, no. 2 (1985): 28–45. It is also described in Larisa V. Shavinina, "The Psychological Essence of the Child Prodigy Phenomenon: Sensitive Periods and Cognitive Experience," *Gifted Child Quarterly* 43, no. 1 (1999): 25–38.

33 **Erwin was exceptionally pale:** G. Révész, *The Psychology of a Musical Prodigy* (1925; reprint, London: Routledge, 1999); Kevin Bazzana, *Lost Genius: The Story of a Forgotten Musical Maverick* (Toronto: McClelland & Stewart, 2007).

33 **While he was still a child:** Erwin's later years, unfortunately, weren't so rosy. For more information on his later life, see Bazzana, *Lost Genius*.

34 **Révész viewed Erwin:** Révész considered whether Erwin was an "infant prodigy" or a "precocious child," but he ultimately rejected both possibilities. Infant prodigies' performances, he wrote, lacked personal inspiration, and their compositions tended to be monotonous, a far cry from Erwin's highly creative, emotional pieces. Nor was Erwin more broadly precocious. Precocious children were advanced in every way, Révész thought, while Erwin was adultlike only in those activities connected with his music.

34 **The next notable contribution:** Franziska Baumgarten, *Wunderkinder: Psychologische Untersuchungen* (Leipzig: Johann Ambrosius Barth, 1930). There is no formal translation of Baumgarten's book. David Feldman very generously shared an unofficial translation created by a former graduate student in the Tufts German department, circa the late 1970s.

34 **Hollingworth's journey with these children:** Details of Hollingworth's initial encounter with Edward come from Charlotte G. Garrison, Agnes Burke, and Leta S. Hollingworth, "The Psychology of a Prodigious Child," *Journal of Applied Psychology* 1, no. 2 (1917): 101–10; Leta S. Hollingworth, Charlotte G. Garrison, and Agnes Burke, "Subsequent History of E——: Five Years After the Initial Report," *Journal of Applied Psychology* 6, no. 2 (1922): 205–10; Leta S. Hollingworth, *Children Above 180 IQ* (Yonkers-on-Hudson, N.Y.: World Book, 1942).

35 **"subnormal intelligence":** Alfred Binet and Th. Simon, *The Development of Intelligence in Children (the Binet-Simon Scale)*, trans. Elizabeth S. Kite (Baltimore: Williams & Wilkins, 1916).

35 **But over time, the test:** Kirk A. Becker, "History of the Stanford-Binet Intelligence Scales: Content and Psychometrics," *Stanford-Binet Intelligence Scales, Fifth Edition, Assessment Service Bulletin No. 1* (2003), http://www.assess.nelson.com/pdf/sb5-asb1.pdf; Deborah L. Ruf, "Use of the SB5 in the Assessment of High Abilities," *Stanford-Binet Intelligence Scales, Fifth Edition, Assessment Service Bulletin No. 3*, http://www.assess.nelson.com/pdf/sb5-asb3.pdf.

35 **In the modern version:** Ruf, "Use of the SB5 in the Assessment of High Abilities."

35 **But the version that Hollingworth used:** Lewis Madison Terman, *The Measurement of Intelligence* (Cambridge, Mass.: Riverside Press, 1916); Ruf, "Use of the SB5 in the Assessment of High Abilities."

35 **Edward's background:** Garrison, Burke, and Hollingworth, "Psychology of a Prodigious Child"; Hollingworth, Garrison, and Burke, "Subsequent History of E——"; Hollingworth, *Children Above 180 IQ.*

35 **"nearly useless to *look* for these children":** Hollingworth, *Children Above 180 IQ,* xiii.

35 **Hollingworth described Edward:** Garrison, Burke, and Hollingworth, "Psychology of a Prodigious Child," 105; Hollingworth, *Children Above 180 IQ,* 142.

35 **she merely meant:** Hollingworth, *Children Above 180 IQ,* 153.

36 **he decided to include:** Feldman, "Mysterious Case of Extreme Giftedness."

36 **Feldman found three children:** These children were later described in Feldman and Goldsmith, *Nature's Gambit.*

36 **He gave them each:** Feldman, "Mysterious Case of Extreme Giftedness"; Feldman, *Beyond Universals in Cognitive Development.*

37 **a more rigorous scientific debate:** For a good description of the debate around this definition, see David Feldman and Martha J. Morelock, "Prodigies and Savants," in *The Cambridge Handbook of Intelligence,* ed. Robert J. Sternberg and Scott Barry Kaufman (Cambridge, U.K.: Cambridge University Press, 2011), 210–34.

37 **"I said it many times":** Feldman, interview, Feb. 21, 2014.

37 **three additional children:** Feldman and Goldsmith, *Nature's Gambit.*

37 **"any theory worth its salt":** Ibid., 109.

37 **The difficulty of using:** Martha J. Morelock, "The Profoundly Gifted Child in Family Context" (Ph.D. diss., Tufts University, 1995).

38 **He went on to earn:** Amy Burkdoll, "Latest Accomplishment: 14-Year-Old Genius Gets His Master's Degree," *Tuscaloosa News,* Aug. 8, 1998; Erica Goode, "The Uneasy Fit of the Precocious and the Average," *New York Times,* March 12, 2002; Brian Dakss, "$1M to Child Prodigy in AOL's 'Gold Rush,'" CBS News, Nov. 10, 2006.

39 **"I think it had some value":** Feldman, interview, Feb. 21, 2014.

39 **Along the way, she wrote up her work:** Joanne Ruthsatz and Douglas K. Detterman, "An Extraordinary Memory: The Case Study of a Musical Prodigy," *Intelligence* 31, no. 6 (2003): 509–18.

40 **"rage to master":** Ellen Winner, "The Rage to Master: The Decisive Role of Talent in the Visual Arts," in *The Road to Excellence: The Acquisition of Expert Performance in the Arts and Sciences, Sports, and Games,* ed. K. Anders Ericsson (Mahwah, N.J.: Lawrence Erlbaum Associates, 1996), 271–301.

40 **Autism, like prodigy:** The CDC provides an overview of autism screening and diagnosis at http://www.cdc.gov/ncbddd/autism/screening.html. To read about a recent interesting effort to identify an autism biomarker, see Tiziano Pramparo et al., "Prediction of Autism by Translation and Immune/

Inflammation Coexpressed Genes in Toddlers from Pediatric Community Practices," *JAMA Psychiatry* 72, no. 4 (2015): 386–94.

40 **Kanner was a psychiatrist:** For information on the history of autism, see Adam Feinstein, *A History of Autism: Conversations with the Pioneers* (Malden, Mass.: Wiley-Blackwell, 2010), and Steve Silberman, *Neurotribes: The Legacy of Autism and the Future of Neurodiversity* (New York: Avery, 2015).

40 **In 1938, Kanner met Donald T.:** Leo Kanner, "Autistic Disturbances of Affective Contact," *Nervous Child* 2 (1943): 217–50.

41 **"extreme autistic aloneness":** Ibid., 242.

41 **Such children were often labeled:** For discussion of former labels for autistic children, see ibid.; Gerald D. Fischbach, "Leo Kanner's 1943 Paper on Autism," Simons Foundation Autism Research Initiative (2007); Roy Richard Grinker, *Unstrange Minds: Remapping the World of Autism* (New York: Basic Books, 2007).

41 **He published a paper:** Kanner, "Autistic Disturbances of Affective Contact." Kanner used the term "early infantile autism" in "Early Infantile Autism," *Journal of Pediatrics* 25, no. 3 (1944): 211–17.

41 **at a children's clinic in Vienna:** Adam Feinstein, *History of Autism;* Uta Frith, "Asperger and His Syndrome," in *Autism and Asperger Syndrome,* ed. Uta Frith (Cambridge, U.K.: Cambridge University Press, 1991), 1–36.

41 **"highly original genius":** Hans Asperger, "'Autistic Psychopathy' in Childhood," in Frith, *Autism and Asperger Syndrome,* 74.

41 **He began using the term:** Feinstein, *History of Autism.*

41 **He called it autistic psychopathy:** There is some wiggle room in this translation. As noted in Frith, *Autism and Asperger Syndrome,* it could also have been translated as "autistic personality disorder" or "autism."

41 **This view is shifting:** Silberman, *Neurotribes.*

41 **"fundamental disorder":** Kanner, "Autistic Disturbances of Affective Contact," 242 (italics removed).

41 **"the shutting-off of relations":** Asperger, "'Autistic Psychopathy' in Childhood," 39.

41 **In the early autism studies:** For a discussion of the fluctuation in autism criteria in the early studies, see Michael Rutter, "Diagnosis and Definition of Childhood Autism," *Journal of Autism and Childhood Schizophrenia* 8, no. 2 (1978): 139–61. Lorna Wing and Judith Gould's classic 1979 paper also has an excellent overview of the difficulty researchers had distinguishing those who had autism from those who did not, as well as trying to sort out whether there are different types of autism. See "Severe Impairments of Social Interaction and Associated Abnormalities in Children: Epidemiology and Classification," *Journal of Autism and Developmental Disorders* 9, no. 1 (1979): 11–29.

42 **"a pseudodiagnostic wastebasket":** Leo Kanner, foreword to *Infantile Autism: The Syndrome and Its Implications for a Neural Theory of Behaviour,* by Bernard Rimland (London: Methuen, 1965), v.

42 **autism has gone from a symptom:** Autism (specifically, "infantile autism") was first listed separately from childhood schizophrenia in the *DSM-III,* published in 1980.

42 **the diagnostic criteria have shifted:** There's a good overview of research on this issue (and a finding that using the diagnostic criteria in *DSM-IV-TR* instead of the criteria in *DSM-III* increased the number of people who qualified as autistic) in Judith S. Miller et al., "Autism Spectrum Disorder Reclassified: A Second Look at the 1980s Utah/UCLA Autism Epidemiologic Study," *Journal of Autism and Developmental Disorders* 43, no. 1 (2013): 200–210.

42 **When work began on the most recent edition of the *DSM*:** David J. Kupfer, Michael B. First, and Darrell A. Regier, eds., *A Research Agenda for DSM-V* (Washington, D.C.: American Psychiatric Association, 2002).

42 **this change was made:** "Autism Spectrum Disorder," *DSM-5 Autism Spectrum Disorder Fact Sheet* (American Psychiatric Association, 2013).

42 **"Behaviour, however reliably":** Uta Frith, "Why We Need Cognitive Explanations of Autism," *Quarterly Journal of Experimental Psychology* 65, no. 11 (2012): 2073–92.

43 **In 2001, he and his colleagues:** Simon Baron-Cohen et al., "The Autism-Spectrum Quotient (AQ): Evidence from Asperger Syndrome/High-Functioning Autism, Males and Females, Scientists and Mathematicians," *Journal of Autism and Developmental Disorders* 31, no. 1 (2001): 5–17.

43 **Autists' family members:** See, for example, Jennifer Gerdts and Raphael Bernier, "The Broader Autism Phenotype and Its Implications on the Etiology and Treatment of Autism Spectrum Disorders," *Autism Research and Treatment* 17 (2011); and A. Pickles et al., "Variable Expression of the Autism Broader Phenotype: Findings from Extended Pedigrees," *Journal of Child Psychology and Psychiatry* 41, no. 4 (2000): 491–502.

44 **AQ results:** Joanne Ruthsatz, "Preliminary Evidence: Expanding the Autistic Spectrum to Include Child Prodigies," in "Behavior Genetics Association 37th Annual Meeting Abstracts," *Behavior Genetics* 37, no. 6 (2007): 734–809.

Chapter 3: The Tiniest Chef

46 **Fellow diners around Manhattan:** The events in this chapter described by Greg Grossman come from a telephone interview conducted on July 12, 2015. The events in this chapter described by Terre Grossman come from telephone interviews conducted on July 19, 2013, May 1 and June 19,

2014, and Sept. 7, 2015; and e-mail. In addition, Terre provided a written summary of Greg's development, a letter she wrote to one of Greg's teachers, essays Greg wrote, menus Greg prepared, documents from Greg's early business ventures, and photographs.

46 **Greg soon began experimenting:** Greg Grossman, "Cooking Autobiography" (writing lab assignment), 2007.

46 **Around the time Greg was nine:** Ibid.

47 **That year, he prepared the meal:** There is some disagreement over whether this first meal was pan-seared scallops or salmon with cucumbers and turnips. The pan-seared scallops descriptions can be found in ibid., and Michelle Trauring, "Greg Grossman: Chef Prodigy," *East Hampton Press and Southampton Press,* Jan. 22, 2013.

48 **Around the time he was twelve:** Greg Grossman, "Introductory Essay" (prepared for Literature and the Law course), 2012; Annie Karni, "A Touch of Classes," *New York Post,* Dec. 19, 2010; Trauring, "Greg Grossman: Chef Prodigy." Some news stories have reported that Greg was even younger than twelve when he began working at the East Hampton restaurant.

48 **The Ross School Café:** Ross School Café menu, http://www.ross.org/menu?rc=0.

48 **When Greg was in fifth grade:** Greg Grossman, telephone interview, July 12, 2015; Greg Grossman, "Introductory Essay."

49 **one day in June:** Information relating to Greg's first catering job comes from a menu he prepared for the event; Barbara Hoffman, "Small Fry," *New York Post,* Aug. 27, 2008; and Greg Grossman, "Introductory Essay."

50 **For the rest of the summer:** Hoffman, "Small Fry"; Barry Gordin, "Gordin's View," *Dan's Papers,* Aug. 29, 2008.

50 **At the James Beard Foundation's:** "Teen Chef Greg Grossman, 13, Demo's Paco Jet at Javit's International Restaurant Show," YouTube video, 1:19, posted by "acookstale," Nov. 3, 2012; "Aspiring Young Chef, a Ross Student, Caters NYC Gala," Hamptons.com, May 8, 2009; "Greg Grossman, Alinea's 13-Year-Old Sous Chef," *Grub Street,* May 18, 2009; "Culinaria Group," *Modern Arts of the Culinary World,* March 11, 2009.

51 **"coming out party":** Andrew Greiner, "Culinary Superstar at Age 13," NBCNewYork.com, May 18, 2009.

51 **Greg demonstrated how to whip up:** Christopher Borrelli, "Greg Grossman: Celebrity Chef Is Just 13," *Chicago Tribune,* May 18, 2009; lobster recipe, 2009 National Restaurant Association demo, May 2009; "Culinaria Group"; "East End's Kiddie Chef Signs Reality-TV Deal," *Grub Street,* June 8, 2009; Maxine Shen, "Lil' Food Dude," *New York Post,* June 9, 2009.

51 **The newspapers ribbed him:** Borrelli, "Greg Grossman: Celebrity Chef Is Just 13"; Shen, "Lil' Food Dude."

51 **When TV came calling:** Borrelli, "Greg Grossman: Celebrity Chef Is Just 13"; "Kid Chef Greg Grossman Failed to Wow VIP Guests with Five-Course Meal," *New York Daily News*, March 15, 2011.

51 **Just one summer after:** "Greg Grossman Thanks Martha Stewart for Ross School Scholarship and Makes Dessert for 450," YouTube video, 17:51, posted by "acookstale," Nov. 3, 2012; Greg Grossman, Twitter entries, July and Aug. 2009.

52 **The conference's host:** Harold McGee, "Modern Cooking & the Erice Workshops on Molecular & Physical Gastronomy," *Curious Cook,* last modified 2011; David Arnold, "What About This?," *Food Arts,* June 2006.

52 **Hervé This:** Patric Kuh, "Proving It," *Gourmet,* Jan. 2005; Sally McGrane, "The Father of Molecular Gastronomy Whips Up a New Formula," *Wired,* July 24, 2007; "The Man Who Unboiled an Egg," *Observer,* Feb. 2008.

53 **Greg hated the term:** Greg was not alone in this sentiment. Others closely associated with molecular gastronomy also clarified their feelings about the term. As Heston Blumenthal explained, "Molecular makes it sound complicated . . . and gastronomy makes it sound elitist." "'Molecular Gastronomy Is Dead.' Heston Speaks Out," *Observer,* Dec. 2006. See also Ferran Adrià et al., "Statement on the 'New Cookery,'" *Observer,* Dec. 9, 2006.

55 **But from Greg's youngest days:** Greg Grossman, "Komputer Kid" (business card); Greg Grossman, "Naturefaces" (overview of Naturefaces idea).

56 **But his need to cook:** Greg Grossman, "Independent Study Proposal: Culinary Technology Lab @Ross," ca. 2009; Greg Grossman, "Culinary Technology Analysis Report (T1)," ca. 2009; Greg, Twitter entries, fall 2009–spring 2010.

56 **Greg and his team prepared:** Vicki Jenkins, e-mail.

57 **It's a score meant to reflect:** "Frequently Asked Questions About the Stanford-Binet Intelligence Scales, Fifth Edition," Thomson Nelson. The memory section of the Stanford-Binet was changed from the fourth edition of the test (which Garrett took) to the fifth edition (which Greg took) so that it focused more on working memory as opposed to short-term memory. The fourth edition also included some working memory components, and all of the other prodigies that Joanne IQ tested were given the fifth edition of the test.

57 **Greg had plenty of amazing-memory anecdotes:** Terre and Ed Grossman to Greg Grossman's teacher, ca. 2003; Greg Grossman, telephone interview, July 12, 2015.

57 **His mind was awash:** Similar anecdotes of extraordinary memory pepper the child prodigy literature. In a case study of a writing prodigy, for example, the authors characterized the child's memory as "extraordinary." He

had excellent recall of his own writing, could recite whole paragraphs of books verbatim at two and a half, and reproduced the exact score of Bach's Sixth Suite from memory at four. See Alan L. Edmunds and Kathryn A. Noel, "Literary Precocity: An Exceptional Case Among Exceptional Cases," *Roeper Review* 25, no. 4 (2003): 185–94.

57 **In a groundbreaking 1946 doctoral dissertation:** Adriaan D. de Groot's work was eventually translated into English. See *Thought and Choice in Chess* (1965; reprint, Amsterdam: Amsterdam University Press, 2008).

57 **Since then, dozens of studies:** For a list of studies on expert memory in various fields, see Kim J. Vicente and JoAnne H. Wang, "An Ecological Theory of Expertise Effects in Memory Recall," *Psychological Review* 105, no. 1 (1998): 33–57.

57 **waiters demonstrate better memories:** Tristan A. Bekinschtein, Julian Cardozo, and Facundo F. Manes, "Strategies of Buenos Aires Waiters to Enhance Memory Capacity in a Real-Life Setting," *Behavioural Neurology* 20, no. 3 (2008): 65–70; H. L. Bennett, "Remembering Drink Orders: The Memory Skill of Cocktail Waitresses," *Human Learning* 2, no. 2 (1983): 157–69; K. Anders Ericsson and Peter G. Polson, "A Cognitive Analysis of Exceptional Memory for Restaurant Orders," in *The Nature of Expertise,* ed. Michelene T. H. Chi, Robert Glaser, and M. J. Farr (Hillsdale, N.J.: Lawrence Erlbaum Associates, 1988), 23–70.

57 **musicians demonstrate better memories:** John A. Sloboda, *The Musical Mind: The Cognitive Psychology of Music* (Oxford: Clarendon Press, 1985); John A. Sloboda, "Visual Perception of Musical Notation: Registering Pitch Symbols in Memory," *Quarterly Journal of Experimental Psychology* 28, no. 1 (1976): 1–16; John A. Sloboda, "Perception of Contour in Music Reading," *Perception* 7, no. 3 (1978): 323–31.

58 **But when chess pieces:** William G. Chase and Herbert A. Simon, "Perception in Chess," *Cognitive Psychology* 4, no. 1 (1973): 55–81.

58 **Psychologists have thus theorized:** K. Anders Ericsson and Peter G. Polson, "An Experimental Analysis of the Mechanisms of a Memory Skill," *Journal of Experimental Psychology: Learning, Memory, and Cognition* 14, no. 2 (1988): 305–16.

58 ***not* superior overall memory:** Chase and Simon, "Perception in Chess"; Vicente and Wang, "Ecological Theory of Expertise Effects in Memory Recall"; K. Anders Ericsson and William G. Chase, "Exceptional Memory," *American Scientist* 70, no. 6 (1982): 607–15.

58 **Again and again, the prodigies:** For an evolutionary explanation of child prodigies tied to working memory, see Larry R. Vandervert, "The Appearance of the Child Prodigy 10,000 Years Ago: An Evolutionary and Developmental Explanation," *Journal of Mind and Behavior* 30, nos. 1 and 2 (2009): 15–32.

59 **"the information tends to be repeated"**: *Diagnostic and Statistical Manual of Mental Disorders,* 4th ed. (Washington, D.C.: American Psychiatric Association, 1994), 68.

59 **"an inordinate number"**: Kanner, "Autistic Disturbances of Affective Contact," 243.

59 **Asperger similarly observed:** Asperger, "'Autistic Psychopathy' in Childhood," 44, 75.

60 **Similar reports of autists:** Tom Fields-Meyer, "What One Father Learned from His Extraordinary Son's Autism," *Atlantic,* Sept. 6, 2011; Gareth Cook, "The Autism Advantage," *New York Times,* Nov. 29, 2012.

60 **But systematic studies:** For overviews of the research on autism and memory, see Jill Boucher, Andrew Mayes, and Sally Bigham, "Memory in Autism Spectrum Disorder," *Psychological Bulletin* 138, no. 3 (2012): 458–96; and Suneeta Kercood et al., "Working Memory and Autism: A Review of Literature," *Research in Autism Spectrum Disorders* 8, no. 10 (2014): 1316–32.

60 **a trait of some:** Andrée-Anne S. Meilleur, Patricia Jelenic, and Laurent Mottron, "Prevalence of Clinically and Empirically Defined Talents and Strengths in Autism," *Journal of Autism and Developmental Disorders* 45, no. 5 (2015): 1354–67.

60 **Nadia:** For the most comprehensive accounts of Nadia, see Lorne Selfe, *Nadia: A Case of Extraordinary Drawing Ability in an Autistic Child* (New York: Academic Press, 1977); and Lorne Selfe, *Nadia Revisited: A Longitudinal Study of an Autistic Savant* (New York: Psychology Press, 2011). For an informative review of the first book, see Nigel Dennis, "Portrait of the Artist," *New York Review of Books,* May 4, 1978.

61 **"island of genius":** Darold A. Treffert, *Islands of Genius: The Bountiful Mind of the Autistic, Acquired, and Sudden Savant* (London: Jessica Kingsley, 2010).

61 **Treffert, a Wiscohsin psychiatrist:** Darold Treffert, telephone interview, April 2, 2015; Darold A. Treffert, "The Savant Registry: A Preliminary Report," Wisconsin Medical Society, https://www.wisconsinmedicalsociety.org.

61 **a varied collection of talents:** For an overview of the savants described here, see Treffert, *Islands of Genius.*

62 **This ability is so predominant:** Treffert, *Islands of Genius.*

62 **"practically every song written":** Bernard Rimland and Deborah Fein, "Special Talents of Autistic Savants," in *The Exceptional Brain: Neuropsychology of Talent and Special Abilities,* ed. Loraine K. Obler and Deborah Fein (New York: Guilford Press, 1988), 479.

62 **John, the first savant:** Treffert, telephone interview, April 2, 2015; Treffert, *Islands of Genius.*

62 **An exceptional memory is:** Darold A. Treffert, "The Savant Syndrome: An Extraordinary Condition," *Philosophical Transactions of the Royal Society B* 364, no. 1522 (2009): 1353.

Chapter 4: Growing a Prodigy

65 **an independent album:** Jonathan Russell, *Stargazer*, CD (2015).
65 **"My brain is in constant music mode":** Jonathan Russell, telephone interview, Jan. 20, 2014.
65 **Music has been a part:** The events in this chapter described by Jonathan Russell come from a telephone interview conducted on June 20, 2014; and e-mail. The events in this chapter described by Eve Weiss come from telephone interviews conducted on Jan. 20 and Nov. 18 and 19, 2014 (with occasional input from Jim Russell); and e-mail. In addition, Eve provided photographs of Jonathan. Jonathan's story was also drawn from his Web site, his YouTube channel (channel ID: jdrcomposer), and news reports, including Marc Ferris, "A Pet Named Mercutio and Spaceship Sheets," *New York Times,* Nov. 17, 2002; Mike Voorheis, "Violin Prodigy, Guitar Expert Return to Jazz Fest," *StarNews,* Feb. 15, 2011; Bobby Caina Calvan, "Sacramento Jazz Jubilee: Youthfull Feats (and Feet)," *Sacramento Bee,* May 27, 2007; Andy Webster, "A Funny Girl Strives to Survive," *New York Times,* April 13, 2011.
65 **His mother, Eve Weiss:** Allen Hughes, "Recital: Eve Weiss, Guitarist," *New York Times,* June 26, 1983.
66 **"teeny tiny violin":** Eve Weiss, telephone interview, Nov. 18, 2014.
69 **"sophisticated improvisations":** Ferris, "Pet Named Mercutio and Spaceship Sheets."
69 **"the rotary phone and the cell phone":** Voorheis, "Violin Prodigy, Guitar Expert Return to Jazz Fest."
70 **"pluckishly improvised":** Calvan, "Sacramento Jazz Jubilee: Youthful Feats (and Feet)."
70 **"sprightly contributions":** Webster, "A Funny Girl Strives to Survive.
71 **Lauren Voiers grew up in Westlake:** The events in this chapter described by Lauren Voiers come from telephone interviews conducted on Feb. 6, Nov. 17, and Dec. 2, 2014, and Sept. 7, 2015; and e-mail. In addition, Lauren provided photographs and images of artwork. The events in this chapter described by Doug Voiers come from a telephone interview conducted on Nov. 30, 2014. The events in this chapter described by Nancy Voiers come from e-mail. Lauren's story was also drawn from her Web site and news reports, including John Soeder, "Cleveland Artist Lauren Voiers Sculpts John Lennon Tribute for Liverpool Park," *Plain Dealer,* Oct. 8, 2010; Ron Vidika, "Lakewood Artist Creates Sculpture Honoring the Beatles' John Lennon,"

Morning Journal, April 10, 2011; Fran Storch, "Two Talented Young Artists Display Their Works at Beck Center Through August 24," *Lakewood Observer,* July 21, 2008; Joe Noga, "Westlake's Lauren Voiers Crafts Sculpture to Honor John Lennon's Legacy of Peace," Cleveland.com, Sept. 29, 2010; "Making the Connection," *ASCENT* (Autumn 2011); "Artist of the Month," *Art List,* May 2007; and Lauren's appearances on local television channels, including Fox 8, WKYC, and ABC Live on 5, many of which can be found on YouTube.

73 **She quickly moved toward cubism:** For an overview of cubism, see "Heilbrunn Timeline of Art History," Metropolitan Museum of Art, http://www.metmuseum.org/toah/.

73 **"a more stained-glass effect":** "Artist of the Month," quoting Lauren Voiers.

75 **"not precocious":** "Training Supermen," *New York Times,* May 7, 1914.

75 **"not geniuses" and "not even exceptionally bright":** Flo Conway and Jim Siegelman, *Dark Hero of the Information Age: In Search of Norbert Wiener* (New York: Basic Books, 2005), 21, quoting a 1909 article from Boston's *Sunday Herald.* For other examples of these parents claiming that their children were not innately exceptional, see "The Boy Prodigy of Harvard," *Current Literature,* March 1910; H. Addington Bruce, "New Ideas in Child Training," *American,* July 1911; H. Addington Bruce, "New Ideas in Child Training," *Journal of Education,* Sept. 21, 1911. On one occasion, though, Wiener did describe Norbert as having a "keen analytical mind" and "tremendous memory." See Conway and Siegelman, *Dark Hero of the Information Age,* 5.

75 **"I could take almost any child":** "Training Supermen."

76 **In their parents' telling:** On the parents' educational philosophies, see "Illustrating a System of Education," *New York Times,* Jan. 7, 1910; "Boy Prodigy of Harvard"; Bruce, "New Ideas in Child Training," *American;* "Give Easy Recipe for Child Prodigies," *New York Times,* Oct. 31, 1920; Greg Daugherty, "The Child Prodigies Who Became 20th-Century Celebrities," *Smithsonian,* June 24, 2013; Winifred Sackville Stoner, *Natural Education* (Indianapolis: Bobbs-Merrill, 1914); Conway and Siegelman, *Dark Hero of the Information Age;* Bruce, "New Ideas in Child Training," *Journal of Education;* and H. Addington Bruce, "Bending the Twig: The Education of the Eleven Year Old Boy Who Lectured Before the Harvard Professors on the Fourth Dimension," *American,* March 1910.

76 **"verses, zoologic and botanic names":** Kanner, "Autistic Disturbances of Affective Contact," 243.

76 **"emotional refrigeration":** Leo Kanner and Leon Eisenberg, "Notes on the Follow-Up Studies of Autistic Children," in *Psychopathology of*

Childhood, ed. Paul H. Hoch and Joseph Zubin (New York: Grune & Stratton, 1955), 229.

76 **He eventually concluded:** Leo Kanner and Leon Eisenberg, "Early Infantile Autism, 1943–1955," *Psychiatric Research Reports of the American Psychiatric Association* 7 (1957): 62. See also Kanner, "Autistic Disturbances of Affective Contact"; Leo Kanner, "Follow-Up Study of Eleven Autistic Children Originally Reported in 1943," *Journal of Autism and Childhood Schizophrenia* 1, no. 2 (1971): 119–45.

76 **Bruno Bettelheim:** Much has been written about Bruno Bettelheim; see, for example, Richard Pollak, *The Creation of Dr. B: A Biography of Bruno Bettelheim* (New York: Simon & Schuster, 1997); and Nina Sutton, *Bettelheim: A Life and a Legacy* (New York: Basic Books, 1996).

77 **"devouring witch":** Bruno Bettelheim, *The Empty Fortress: Infantile Autism and the Birth of the Self* (New York: Free Press, 1967), 71.

77 **In the first of these:** Susan Folstein and Michael Rutter, "Infantile Autism: A Genetic Study of 21 Twin Pairs," *Journal of Child Psychology and Psychiatry* 18, no. 4 (1977): 297–321.

77 **These findings were buttressed:** Edward R. Ritvo et al., "Concordance for the Syndrome of Autism in 40 Pairs of Afflicted Twins," *American Journal of Psychiatry* 142, no. 1 (1985): 74–77; Suzanne Steffenburg et al., "A Twin Study of Autism in Denmark, Finland, Iceland, Norway and Sweden," *Journal of Child Psychology and Psychiatry and Allied Disciplines* 30, no. 3 (1989): 405–16; and A. Bailey et al., "Autism as a Strongly Genetic Disorder: Evidence from a British Twin Study," *Psychological Medicine* 25, no. 1 (1995): 63–77. For a more recent twin study, see Rebecca E. Rosenberg et al., "Characteristics and Concordance of Autism Spectrum Disorders Among 277 Twin Pairs," *Archives of Pediatrics and Adolescent Medicine* 163, no. 10 (2009): 907–14. For a study finding a higher-than-expected incidence of concordance for autism in fraternal twins, see Joachim Hallmayer et al., "Genetic Heritability and Shared Environmental Factors Among Twin Pairs with Autism," *Archives of General Psychiatry* 68, no. 11 (2011): 1095–102.

77 **studies that identified a higher prevalence:** Early studies focused on an elevated rate of intellectual disability among autists' family members; for a review, see Patrick Bolton and Michael Rutter, "Genetic Influences in Autism," *International Review of Psychiatry* 2, no. 1 (1990): 67–80. For an example of an early study identifying subclinical social and communication deficits in autists' families, see P. Bolton, "A Case-Control Family History Study of Autism," *Journal of Child Psychology and Psychiatry* 35, no. 5 (1994): 877–900. For a more recent review, see Gerdts and Bernier, "Broader Autism Phenotype and Its Implications on the Etiology and Treatment of Autism Spectrum Disorders."

77 **"striking and extreme"**: Feldman and Goldsmith, *Nature's Gambit,* 43.

77 **Kelvin Doe, for example:** THNKR, "Kelvin Doe at TEDxTeen"; Kelvin Doe, "Persistent Experimentation," TEDxTeen; Ian Steadman, "Teenager Signs £60,000 Contract to Develop His Own Solar Panel Technology," *Wired UK,* Oct. 10, 2013; Oswald Hanciles, "The Mystery (and Challenge) of Kelvin Doe," *Concord Times,* July 1, 2015; Nina Strochlic, "Why the Clintons Love Sierra Leone's Boy Genius," *Daily Beast,* Sept. 26, 2013.

78 **"a beautifully choreographed"**: Feldman and Goldsmith, *Nature's Gambit,* 14.

78 **"stand out like Gulliver"**: Goode, "Uneasy Fit of the Precocious and the Average."

78 **the music prodigy:** "Musical Prodigy, Bluejay," *60 Minutes,* Nov. 30, 2006.

79 **"real learning"**: Written summary of Alex's and William's development, provided by Lucie, 2012.

79 **"He then put his pencil down"**: Ibid.

80 **"I think we'll just avoid"**: Lucie, telephone interview, Sept. 12, 2014.

81 **"rage to master"**: Winner, "The Rage to Master: The Decisive Role of Talent in the Visual Arts."

Chapter 5: The Evidence Mounts

82 **He was born in Newport-on-Tay:** The events in this chapter described by Mike Wawro come from telephone interviews conducted on Dec. 14 and 16, 2014, and March 30, 2015; and e-mail. In addition, Mike provided photographs, images of Richard's artwork, and a record of Richard's drawings and sales. The events described by Laurence A. Becker come from a telephone interview conducted on Dec. 17, 2014; and e-mail. Richard's story was also drawn from his Web site; documentaries, including Ron Zimmerman and Laurence A. Becker, *With Eyes Wide Open* (1983); and news reports, including "A Conversation with Laurence A. Becker: On the Gifted Handicapped," *The Human Condition,* adapted for publication by Charlene Warren (Austin: Hogg Foundation for Mental Health and the University of Texas, 1981); Ann Shearer, "Subnormal—but an Astounding Artist," *Guardian,* April 18, 1970; "Richard Wawro," *Scotsman,* March 9, 2006; "Richard Wawro," *Telegraph,* March 11, 2006; Jonathan Brinckman, "Autistic Artist Can't Explain How His Talent Works," *New Haven Register,* May 20, 1990; Lee Kelly, "Austin Exhibit of Wawro's Artwork Kicks Off National Tour," *Austin American-Statesman,* April 8, 1990; Mary Meehan, "Blind Artist Shows His True Colors," *Orlando Sentinel,* May 10, 1990; Adam Geller, "Art That Speaks—Autistic and Nearly Blind, Scot Inspires Disabled Kids," *Record,* April 7, 1998; Marian Bohusz-Szyszko, "Phenomenon," ca. mid-1960s, translated version used by the Wawro family provided by Mike Wawro; Joseph Blank, "I Can See Feeling Good,"

Reader's Digest, Nov. 1983; and Denise Gamino, "Renowned Artist Savant Visits Austin," *Austin American-Statesman,* Oct. 13, 1994.

82 **"It was enough to drive you mad"**: Brinckman, "Autistic Artist Can't Explain How His Talent Works."

83 **"was never still"**: Molly Leishman, "Richard Wawro, Artist," http://www.wawro.net/Richard_Wawro/Early_life.html.

84 **"It wasn't the usual picture"**: Molly Leishman, television interview, Zimmerman and Becker, *With Eyes Wide Open.*

85 **"thunderstruck"**: Bohusz-Szyszko, "Phenomenon."

86 **Richard Demarco**: "Richard Wawro," *Scotsman.*

86 **"On the first evening"**: Blank, "I Can See Feeling Good."

87 **"hold them up"**: "A Conversation with Laurence A. Becker: On the Gifted Handicapped"; this publication includes the bookstore manager quotation from Richard A. Abrams, *American Statesman,* ca. 1980.

88 **At some of his exhibitions**: For an excellent overview of Richard's technique, see Zimmerman and Becker, *With Eyes Wide Open.*

88 **Richard showed his parents**: To see a recording of this celebration ritual, see ibid.

89 **"I can't get over the number"**: Ibid. (narrator Cactus Pryor is speaking).

89 **Autism became the official**: See, for example, "Richard Wawro," *Scotsman.*

90 **Olive had contracted rubella**: For an overview of studies investigating a potential relationship between congenital rubella and autism, see Jane E. Libbey et al., "Autistic Disorder and Viral Infections," *Journal of Neuro-Virology* 11, no. 1 (2005): 1–10.

90 **at his brother's flat in Glasgow**: Zimmerman and Becker, *With Eyes Wide Open.*

91 **"world champion picture"**: Brinckman, "Autistic Artist Can't Explain How His Talent Works."

94 **"He talks of little else"**: Kanner, "Autistic Disturbances of Affective Contact," 233.

94 **Asperger, too, noted a tendency**: Asperger, "'Autistic Psychopathy' in Childhood," 72.

95 **"highly restricted, fixated interests"**: In the *DSM-5,* this description falls under the umbrella category of "restricted, repetitive patterns of behavior, interests, or activities." *Diagnostic and Statistical Manual of Mental Disorders,* fifth ed. (Washington, D.C.: American Psychiatric Association, 2013).

95 **Many autists demonstrate such circumscribed interests**: See, for example, Lauren M. Turner-Brown et al., "Phenomenology and Measurement of Circumscribed Interests in Autism Spectrum Disorders," *Autism* 15, no. 4 (2011): 437–56; Mikle South, Sally Ozonoff, and William M. McMahon,

"Repetitive Behavior Profiles in Asperger Syndrome and High-Functioning Autism," *Journal of Autism and Developmental Disorders* 35, no. 2 (2005): 145–58; and Turner-Brown et al., "Phenomenology and Measurement of Circumscribed Interests in Autism Spectrum Disorders."

95 **Several researchers have observed:** Turner-Brown et al., "Phenomenology and Measurement of Circumscribed Interests in Autism Spectrum Disorders"; Ami Klin et al., "Circumscribed Interests in Higher Functioning Individuals with Autism Spectrum Disorders: An Exploratory Study," *Research & Practice for Persons with Severe Disabilities* 32, no. 2 (2007): 89–100.

95 **the autists' interests can lie in any area:** See, for example, South, Ozonoff, and McMahon, "Repetitive Behavior Profiles in Asperger Syndrome and High-Functioning Autism."

95 **From the autist's family's perspective:** Céline Mercier, Laurent Mottron, and Sylvie Belleville, "A Psychosocial Study on Restricted Interests in High Functioning Persons with Pervasive Developmental Disorders," *Autism* 4, no. 4 (2000): 406–25; South, Ozonoff, and McMahon, "Repetitive Behavior Profiles in Asperger Syndrome and High-Functioning Autism."

96 **Joanne took stock:** Joanne Ruthsatz and Jourdan B. Urbach, "Child Prodigy: A Novel Cognitive Profile Places Elevated General Intelligence, Exceptional Working Memory, and Attention to Detail at the Root of Prodigiousness," *Intelligence* 40, no. 5 (2012): 419–26.

96 **Autism is more common among men:** Eric Fombonne, "Epidemiology of Pervasive Developmental Disorders," *Pediatric Research* 65, no. 6 (2009): 591–98.

96 **The reasons for this asymmetry:** See, for example, Bonnie Auyeung et al., "Fetal Testosterone and Autistic Traits," *British Journal of Psychology* 100, no. 1 (2009): 1–22.

96 **"extreme male brain":** See, for example, Simon Baron-Cohen, "The Extreme Male Brain Theory of Autism," *TRENDS in Cognitive Sciences* 6, no. 6 (2002): 248–54. A similar idea was proposed by Hans Asperger. See Asperger, "'Autistic Psychopathy' in Childhood," 84.

96 **Others have proposed:** S. Jacquemont et al., "A Higher Mutational Burden in Females Supports a 'Female Protective Model' in Neurodevelopmental Disorders," *American Journal of Human Genetics* 94, no. 3 (2014): 415–25; Elise B. Robinson et al., "Examining and Interpreting the Female Protective Effect Against Autistic Behavior," *Proceedings of the National Academy of Sciences* 110, no. 13 (2013): 5258–62; T. W. Frazier et al., "Behavioral and Cognitive Characteristics of Females and Males with Autism in the Simons Simplex Collection," *Journal of the American Academy of Child and Adolescent Psychiatry* 53, no. 3 (2014): 329–40.

97 **Still others suggest:** Alexandra M. Head, Jane A. McGillivray, and Mark A. Stokes, "Gender Differences in Emotionality and Sociability in Children with Autism Spectrum Disorders," *Molecular Autism* 5, no. 19 (2014); T. W. Frazier et al., "Behavioral and Cognitive Characteristics of Females and Males with Autism in the Simons Simplex Collection."

97 **"a universal feature":** Simon Baron-Cohen et al., "Talent in Autism: Hyper-Systemizing, Hyper-Attention to Detail, and Sensory Hypersensitivity," *Philosophical Transactions of the Royal Society B* 364, no. 1522 (2009): 1377.

97 **Heightened attention:** See, for example, Francesca Happé and Uta Frith, "The Weak Coherence Account: Detail-Focused Cognitive Style in Autism Spectrum Disorders," *Journal of Autism and Developmental Disorders* 36, no. 1 (2006): 5–25.

97 **In one highly publicized 1996 incident:** Rick Bragg, "Autism No Handicap, Boy Defies Swamp," *New York Times,* Aug. 17, 1996; Patrick Rogers, "Alive!," *People,* Sept. 2, 1996; "Autistic Boy Is Found After 4 Nights in Swamp," Associated Press, Aug. 12, 1996.

97 **Afterward, Stephen reproduced it:** "The Foolish Wise Ones," *QED*, 1986.

98 **On other occasions, he has produced:** *Beautiful Minds: A Voyage into the Brain,* 2006; Treffert, *Islands of Genius.*

98 **They outscored those:** Ruthsatz and Urbach, "Child Prodigy."

98 **Jonathan Russell, for example:** Jonathan Russell, telephone interview, Jan. 20, 2014.

Chapter 6: Chromosome 1

99 **Kristine Barnett's first pregnancy:** The events in this chapter described by Kristine Barnett come from telephone interviews conducted on Dec. 22, 2014 (with occasional input from Jacob Barnett), and Oct. 13, 2015. The events in this chapter described by Becky Pearson come from a telephone interview conducted on May 29, 2015. Jacob's story was also drawn from Kristine Barnett's book, *The Spark: A Mother's Story of Nurturing, Genius, and Autism* (New York: Random House, 2013), Kindle edition; Jacob's TEDxTeen talk, "Forget What You Know"; and news reports, including "Groupon, Qatar, Jake Barnett," *60 Minutes,* Jan. 15, 2012; Dan McFeely, "Genius at Work: 12-Year-Old Is Studying at IUPUI," *Indianapolis Star,* March 20, 2011; Paul Wells, "Jacob Barnett, Boy Genius," *Maclean's,* Sept. 1, 2013; Louise Carpenter, "When Jacob Barnett Was 3, His Mother Was Told That He Would Never Be Able to Read," *Times (London),* April 20, 2013; Eric Weddle, "Boy Genius' Celebrity Grows with New Book, Movie Deal," *Indianapolis Star,* April 8, 2013; "13-Year-Old Beats the Odds," *Frankfort (Ind.) Times,* Jan. 31, 2011; and Jacob's and his family's appearances on *Glenn Beck, The Agenda with Steve Paikin,* and BBC World News.

104 **"Night-night, baby bagel":** Kristine Barnett, *Spark,* 53.
105 **"Some of these things":** *The Agenda with Steve Paikin.*
107 **"kindergarten boot camp":** Kristine Barnett, *Spark,* 67, 80.
108 **"he looked at me reproachfully":** Ibid., 119.
111 **Sometimes numbers or words:** Noam Sagiv et al., "What Is the Relationship Between Synaesthesia and Visuo-spatial Number Forms?," *Cognition* 101, no. 1 (2006): 114–28; Jamie Ward, Julia Simner, and Vivian Auyeung, "A Comparison of Lexical-Gustatory and Grapheme-Colour Synaesthesia," *Cognitive Neuropsychology* 22, no. 1 (2005): 28–41.
111 **months occupy a spatial location:** Daniel Smilek et al., "Ovals of Time: Time-Space Associations in Synaesthesia," *Consciousness and Cognition* 16, no. 2 (2007): 507–19.
111 **One woman perceives August:** Julia Simner and Emma Holenstein, "Ordinal Linguistic Personification as a Variant of Synesthesia," *Journal of Cognitive Neuroscience* 19, no. 4 (2007): 694–703.
111 **The most common form:** Julia Simner et al., "Synaesthesia: The Prevalence of Atypical Cross-Modal Experiences," *Perception* 35, no. 8 (2006): 1024–33; A. N. Rich, J. L. Bradshaw, and J. B. Mattingley, "A Systematic, Large-Scale Study of Synaesthesia: Implications for the Role of Early Experience in Lexical-Colour Associations," *Cognition* 98, no. 1 (2005): 53–84.
111 **Researchers have picked up:** Kylie J. Barnett et al., "Familial Patterns and the Origins of Individual Differences in Synaesthesia," *Cognition* 106, no. 2 (2008): 871–93.
112 **Tests taken months:** See, for example, Caroline Yaro and Jamie Ward, "Searching for Shereshevskii: What Is Superior About the Memory of Synaesthetes?," *Quarterly Journal of Experimental Psychology* 60, no. 5 (2007): 681–95.
112 **Synesthesia is particularly interesting:** See, for example, Julia Simner, Neil Mayo, and Mary-Jane Spiller, "A Foundation for Savantism? Visuo-spatial Synaesthetes Present with Cognitive Benefits," *Cortex* 45, no. 10 (2009): 1246–60.
112 **"Every number or math problem":** "Groupon, Qatar, Jake Barnett."
112 **Over the years:** See, for example, Laura Cesaroni and Malcolm Garber, "Exploring the Experience of Autism Through Firsthand Accounts," *Journal of Autism and Developmental Disorders* 21, no. 3 (1991): 303–13.
112 **He also has Asperger's disorder:** Simon Baron-Cohen et al., "Savant Memory in a Man with Colour Form-Number Synaesthesia and Asperger Syndrome," *Journal of Consciousness Studies* 14, nos. 9–10 (2007): 237–51.
112 **a recent study found:** Simon Baron-Cohen et al., "Is Synaesthesia More Common in Autism?," *Molecular Autism* 4, no. 40 (2013). See also Julian E. Asher et al., "A Whole-Genome Scan and Fine-Mapping Linkage Study of Auditory-Visual Synaesthesia Reveals Evidence of Linkage to

Chromosomes 2q24, 5q33, 6p12, and 12p12," *American Journal of Human Genetics* 84, no. 2 (2009): 279–85.

113 **wrote a story about Jacob:** "13-Year-Old Beats the Odds."

113 **Two months later:** McFeely, "Genius at Work."

115 **extracted the DNA:** For information on the technique used to collect and extract the DNA, see Michael R. Goode et al., "Collection and Extraction of Saliva DNA for Next Generation Sequencing," *Journal of Visualized Experiments* 90 (2014).

115 **"a big text file":** Chris Bartlett, telephone interview, April 21, 2015.

115 **sought-after genetic mutation:** For a discussion of the challenges in finding common language to describe genetic variants, see Roshan Karki et al., "Defining 'Mutation' and 'Polymorphism' in the Era of Personal Genomics," *BMC Medical Genomics* 8, no. 37 (2015). We use the term "mutation" here to mean a change in DNA such that, as the National Library of Medicine's *Genetics Home Reference* puts it, "the sequence differs from what is found in most people."

116 **There was also one clear hit:** Joanne Ruthsatz et al., "Molecular Genetic Evidence for Shared Etiology of Autism and Prodigy," *Human Heredity* 79, no. 2 (2015): 53–59.

116 **"We did gamble":** Chris Bartlett, telephone interview, Nov. 24, 2014.

117 **The short arm of chromosome 1:** Neil Risch et al., "A Genomic Screen of Autism: Evidence for a Multilocus Etiology," *American Journal of Human Genetics* 65, no. 2 (1999): 493–507; Mari Auranen et al., "A Genomewide Screen for Autism-Spectrum Disorders: Evidence for a Major Susceptibility Locus on Chromosome 3q25-27," *American Journal of Human Genetics* 71, no. 4 (2002): 777–90; Mari Auranen et al., "Analysis of Autism Susceptibility Gene Loci on Chromosomes 1p, 4p, 6q, 7q, 13q, 15q, 16p, 17q, 19q, and 22q in Finnish Multiplex Families," *Molecular Psychiatry* 5, no. 3 (2000): 320–22.

117 **perhaps it's because there aren't many people:** A couple of studies have examined the genetic roots of savant skills. See Erika L. Nurmi et al., "Exploratory Subsetting of Autism Families Based on Savant Skills Improves Evidence of Genetic Linkage to 15q11-q13," *Child and Adolescent Psychiatry* 42, no. 7 (2003): 856–63; and D. Q. Ma et al., "Ordered-Subset Analysis of Savant Skills in Autism for 15q11-q13," *American Journal of Medical Genetics* 135B, no. 1 (2005): 38–41.

Chapter 7: The Empathy Puzzle

118 **In the late 1970s:** Uta Frith, telephone interview, Aug. 4, 2015.

118 **As Frith recalls:** Ibid.; Frith, "Why We Need Cognitive Explanations of Autism."

118   **some studies put the figure:** M. K. DeMyer, "The Measured Intelligence of Autistic Children," *Journal of Autism and Childhood Schizophrenia* 4, no. 1 (1974): 42–60. See also Uta Frith, *Autism: Explaining the Enigma* (Cambridge, Mass.: Basil Blackwell, 1989), 53–54, for a review of several studies that examined the relationship between autism and IQ.

118   **This relatively narrow perception:** For a review of the hunt for a cognitive explanation for autism, see Uta Frith, "Why We Need Cognitive Explanations of Autism."

119   **Most assumed:** See, for example, Simon Baron-Cohen, "The Autistic Child's Theory of Mind: A Case of Specific Developmental Delay," *Journal of Child Psychology and Psychiatry* 30, no. 2 (1989): 285–98; and Uta Frith, John Morton, and Alan M. Leslie, "The Cognitive Basis of a Biological Disorder: Autism," *Trends in Neurosciences* 14, no. 10 (1991): 433–38.

119   **The executive function theory:** For an example of a relatively early paper on the executive function theory, see Sally Ozonoff, Bruce F. Pennington, and Sally J. Rogers, "Executive Function Deficits in High-Functioning Autistic Individuals: Relationship to Theory of Mind," *Journal of Child Psychology and Psychiatry* 32, no. 7 (1991): 1081–105.

119   **According to the weak central coherence theory:** For early works on weak central coherence, see Frith, *Autism;* and Uta Frith and Francesca Happé, "Autism: Beyond 'Theory of Mind,'" *Cognition* 50, nos. 1–3 (1994): 115–32.

119   **"peculiar pattern":** Frith, *Autism,* 174.

119   **In the 1970s, two researchers:** David Premack and Guy Woodruff, "Does the Chimpanzee Have a Theory of Mind?," *Behavioral and Brain Sciences* 1 (1978): 515–26.

120   **In the early 1980s, two researchers:** Heinz Wimmer and Josef Perner, "Beliefs About Beliefs: Representation and Constraining Function of Wrong Beliefs in Young Children's Understanding of Deception," *Cognition* 13, no. 1 (1983): 103–28.

120   **To find out, Baron-Cohen:** Simon Baron-Cohen, Alan M. Leslie, and Uta Frith, "Does the Autistic Child Have a 'Theory of Mind'?," *Cognition* 21, no. 1 (1985): 37–46. For more on the history of this experiment, see Frith, "Why We Need Cognitive Explanations of Autism."

121   **Theory of mind research proliferated:** See, for example, Simon Baron-Cohen et al., "Recognition of Faux Pas by Normally Developing Children and Children with Asperger Syndrome or High-Functioning Autism," *Journal of Autism and Developmental Disorders* 29, no. 5 (1999): 407–18; and Simon Baron-Cohen et al., "The 'Reading the Mind in the Eyes' Test Revised Version: A Study with Normal Adults, and Adults with Asperger Syndrome or High-Functioning Autism," *Journal of Child Psychology and Psychiatry* 42, no. 2 (2001): 241–52.

121 **A few autism researchers had noted:** See, for example, Michael Rutter, David Greenfeld, and Linda Lockyer, "A Five to Fifteen Year Follow-Up Study of Infantile Psychosis," *British Journal of Psychiatry* 113 (1967): 1183–99, 1187; and Leon Eisenberg, "The Autistic Child in Adolescence," *American Journal of Psychiatry* 112, no. 8 (1956): 607–12, 611.

121 **"deficits in the normal process":** Simon Baron-Cohen, "The Cognitive Neuroscience of Autism," *Journal of Neurology, Neuroscience, and Psychiatry* 75, no. 7 (2004): 945–48.

121 **In piloting this test:** Simon Baron-Cohen and Sally Wheelwright, "The Empathy Quotient: An Investigation of Adults with Asperger Syndrome or High Functioning Autism, and Normal Sex Differences," *Journal of Autism and Developmental Disorders* 34, no. 2 (April 2004): 163–75.

121 **"one of the key characteristics":** Pilar Rueda, Pablo Fernández-Berrocal, and Kimberly A. Schonert-Reichl, "Empathic Abilities and Theory of Mind in Adolescents with Asperger Syndrome: Insights from the Twenty-first Century," *Review Journal of Autism and Developmental Disorders* 1, no. 4 (2014): 327–43.

121 **"marked by empathy deficits":** Jean Decety and Meghan Meyer, "From Emotion Resonance to Empathic Understanding: A Social Developmental Neuroscience Account," *Development and Psychopathology* 20, no. 4 (2008): 1053–80, 1053.

122 **The first of these careers:** The events in this chapter described by Jourdan Urbach come from an interview conducted on July 2, 2014. The events in this chapter described by Eric Drier come from a telephone interview conducted on June 4, 2015. Jourdan's story was also drawn from his TEDxYSE talk, "The Children of the Tenth Floor," his Web site, concert posters, event notices, and news reports, including Marcelle S. Fischler, "At 11, a Violin Virtuoso and Author, Too," *New York Times,* March 9, 2003; Gina Salamone, "Prodigy Jourdan Urbach Is Maestro of Kids' Charity Drive," *New York Daily News,* Oct. 17, 2007; Lauren LaCapra, "Fiddle Prodigy Tops MS Show," *New York Daily News,* Jan. 4, 2005; "Jourdan Urbach to Be Featured with the Park Avenue Chamber Symphony at Carnegie Hall, 10/27," *Broadway World,* Aug. 1, 2013; Rochelle Kraut, "Jourdan Urbach: Our Youngest Researcher," *Inside MS,* June–July 2007; Susan Jackson, "Q&A with Jourdan Urbach," *Juilliard Journal,* Feb. 2011; Ula Ilnytzky, "Jefferson Awards 2012: Jourdan Urbach, Violin Prodigy, Among Recipients," *Huffington Post,* March 6, 2012; "Jourdan Urbach," YouTube video, 5:52, appearance on WTNH 8 *Connecticut Style,* posted by "CT STYLE," June 24, 2011; Jessica Guenzel, "One for the Books," *Newsday,* Feb. 9, 2003; "Urbach Charity Performances," *Roslyn News,* Nov. 3, 2006; Rita Delfiner, "Violanthropist," *New York Post,* Aug. 11, 2008; Damian Da Costa, "With Prodigy Urbach, Park

Avenue Chamber Symphony Soared Saturday," *New York Observer,* Nov. 24, 2008; Barbara Hoffman, "Fiddling for a Cause," *New York Post,* Nov. 22, 2008; "Violin Virtuoso Jourdan Urbach Headlines April Benefit Concert at the Shubert," *Stamford Plus,* Feb. 27, 2010; Britt Hysen, "Prodigy Violinist Jourdan Urbach Launches Video App," *Millennial,* June 27, 2014; "More Intel Semifinalists from RHS," *Roslyn News,* Jan. 23, 2009; "Way to Go! Jourdan Urbach of Roslyn High School," *Newsday,* July 4, 2009; "The Next Wave," *People,* April 12, 2004; Cynthia Daniels, "'A Renaissance Boy,'" *Newsday,* Jan. 17, 2005; Corey Kilgannon, "With Dues-Paying Years Over, These Musicians Are Studying at Juilliard," *New York Times,* Jan. 23, 2005; Rahel Musleah, "Jourdan Urbach, Raising Money for Kids' Health," *Family Circle,* Feb. 2006; Jim Shelton, "Yale's Jourdan Urbach Raises Millions for Children's Charities," *New Haven Register,* Dec. 15, 2010; "The 2007 Liberty Medals," *New York Post,* Oct. 17, 2007; Jennifer Fermino, "Gala Salutes a Perfect 10," *New York Post,* Oct. 18, 2007; Kristen Mascia, "Twenty Teens Who Will Change the World"; and Jourdan's television appearances, including on *CBS Sunday Morning, Lou Dobbs Tonight, Good Morning America, Today,* and *Inside Edition.*

122 **"the tiniest violin"**: Salamone, "Prodigy Jourdan Urbach Is Maestro of Kids' Charity Drive."

126 **"dramatically illustrated"**: Ilnytzky, "Jefferson Awards 2012," quoting Arney Rosenblat.

128 **In 2002, Tania Barkat:** At the time, her name was Tania Rinaldi. The details of Tania's work come from telephone interviews with Tania Barkat conducted on June 4 and Sept. 11, 2015; and Maia Szalavitz, "The Boy Whose Brain Could Unlock Autism," *Matter,* Dec. 11, 2013.

128 **Human Brain Project:** Mark Honigsbaum, "Human Brain Project: Henry Markram Plans to Spend €1bn Building a Perfect Model of the Human Brain," *Guardian,* Oct. 15, 2013; Tim Requarth, "Bringing a Virtual Brain to Life," *New York Times,* March 18, 2013.

129 **It turned out that the brains:** For more on this research and some of the follow-up studies, see Henry Markram, Tania Rinaldi, and Kamila Markram, "The Intense World Syndrome—an Alternative Hypothesis for Autism," *Frontiers in Neuroscience* 1, no. 1 (2007); Tania Rinaldi et al., "Elevated NMDA Receptor Levels and Enhanced Postsynaptic Long-Term Potentiation Induced by Prenatal Exposure to Valproic Acid," *Proceedings of the National Academy of Sciences* 104, no. 33 (2007): 13501–6; Tania Rinaldi, Filad Silberberg, and Henry Markram, "Hyperconnectivity of Local Neocortical Microcircuitry Induced by Prenatal Exposure to Valproic Acid," *Cerebral Cortex* 18, no. 4 (2008): 763–70; Kamila Markram et al., "Abnormal Fear Conditioning and Amygdala Processing in an Animal Model of

Autism," *Neuropsychopharmacology* 33, no. 4 (2008): 901–12; Kamila Markram and Henry Markram, "The Intense World Theory—a Unifying Theory of the Neurobiology of Autism," *Frontiers in Human Neuroscience* 4, no. 224 (2010); Mônica R. Favre et al., "Predictable Enriched Environment Prevents Development of Hyper-Emotionality in the VPA Rat Model of Autism," *Frontiers in Neuroscience* 9, no. 127 (2015).

130 **It's yet to be proved:** Szalavitz, "Boy Whose Brain Could Unlock Autism"; Anna Remington and Uta Frith, "Intense World Theory Raises Intense Worries," *Spectrum,* Jan. 21, 2014.

130 **"emotional barometer":** "Autism, Empathy, and Understanding," *MOM— Not Otherwise Specified,* Oct. 28, 2013, comment by Jennifer, posted on Oct. 28, 2013.

130 **"mood ring":** Ibid., comment by Bailzebub, posted on Oct. 29, 2013. See also Liane Kupferberg Carter, "Autism and Empathy," *Huffington Post,* May 17, 2013.

130 **"emotional tornadoes":** John Scott Holman, "Interview: Henry and Kamila Markram About the Intense World Theory for Autism," Wrong-Planet.net, Jan. 6, 2012, comment by SpatialEd, Jan. 7, 2012 (comments have been removed from site).

130 **a woman felt as if:** "Autism, Empathy, and Understanding," *MOM—Not Otherwise Specified,* Oct. 28, 2013, comment by Aimee, Nov. 7, 2013.

130 **"like an exposed nerve":** Holman, "Interview: Henry and Kamila Markram About the Intense World Theory for Autism," comment by AsteroidNap, Jan. 10, 2012 (comments have been removed from site).

130 **"go into sensory lock down":** Holman, "Interview: Henry and Kamila Markram About the Intense World Theory for Autism," comment by Awiddershinlife, Jan. 6, 2012 (comments have been removed from site).

131 **As one individual put it:** Holman, "Interview: Henry and Kamila Markram About the Intense World Theory for Autism," comment by Matt1988, Jan. 11, 2012 (comments have been removed from site).

131 **"unquestionably endowed":** Kanner, "Autistic Disturbances of Affective Contact," 247.

131 **"a high level of original thought":** Asperger, "'Autistic Psychopathy' in Childhood," 37.

131 **A late 1970s study:** Bernard Rimland, "Inside the Mind of the Autistic Savant," *Psychology Today,* Aug. 1978, 69–80. For an academic account of this work, see Rimland and Fein, "Special Talents of Autistic Savants." For a more recent, higher estimate of the prevalence of savant skills among autists, see Meilleur, "Prevalence of Clinically and Empirically Defined Talents and Strengths in Autism."

132 **Her 1991 translation:** Asperger, "'Autistic Psychopathy' in Childhood." A detailed account of Asperger's work was written by Lorna Wing and

published in 1981. See "Asperger's Syndrome: A Clinical Account," *Psychological Medicine* 11, no. 1 (1981): 115–29.

132 **"Able autistic individuals":** Asperger, "'Autistic Psychopathy' in Childhood," 88–89.

133 **But some researchers questioned:** See Wing, "Asperger's Syndrome"; Feinstein, *History of Autism*.

133 **Even before the *DSM-5* was issued:** For an overview, see Miller et al., "Autism Spectrum Disorder Reclassified." See also Silberman, *Neurotribes*.

133 **A 2006 review study:** See Meredyth Goldberg Edelson, "Are the Majority of Children with Autism Mentally Retarded? A Systematic Evaluation of the Data," *Focus on Autism and Other Developmental Disabilities* 21, no. 2 (2006): 66–83. For additional discussion of this issue, see Markram and Markram, "Intense World Theory."

133 **Another study found:** Tony Charman et al., "IQ in Children with Autism Spectrum Disorders: Data from the Special Needs and Autism Project (SNAP)," *Psychological Medicine* 41, no. 3 (2011): 619–27.

133 **Another line of research:** Michelle Dawson et al., "The Level and Nature of Autistic Intelligence," *Psychological Science* 18, no. 8 (2007): 657–62. See also Sven Bölte, Isabel Dziobek, and Fritz Poustka, "Brief Report: The Level and Nature of Autistic Intelligence Revisited," *Journal of Autism and Developmental Disorders* 39, no. 4 (2009): 678–82.

133 **The executive function theory:** For reviews of the literature on executive dysfunction, see Elizabeth Pellicano, "The Development of Executive Function in Autism," *Autism Research and Treatment* 2012, Article ID 146132 (2012); and Elisabeth L. Hill, "Evaluating the Theory of Executive Dysfunction in Autism," *Developmental Review* 24, no. 2 (2004): 189–233.

134 **The weak central coherence theory:** See, for example, Happé and Frith, "Weak Coherence Account."

134 **The new theory emphasized:** Simon Baron-Cohen, "Autism: The Empathizing-Systemizing (E-S) Theory," *Year in Cognitive Neuroscience* 1156 (2009): 68–80. For an excellent overview of the evolution of these theories from an insider's perspective, see Frith, "Why We Need Cognitive Explanations of Autism."

134 **These researchers think:** For a review of this research, particularly that focused on individuals diagnosed with Asperger's syndrome, see Rueda, Fernández-Berrocal, and Schonert-Reichl, "Empathic Abilities and Theory of Mind in Adolescents with Asperger Syndrome." For more information on Simon Baron-Cohen's perspective on the cognitive/affective empathy divide, see his TEDxHousesofParliament talk, "The Erosion of Empathy," and his book, *The Science of Evil: On Empathy and the Origins of Cruelty* (New York: Basic Books, 2011).

134 **The drive to systematize:** Baron-Cohen, *Science of Evil;* Simon Baron-Cohen, "Autism—'Autos': Literally, a Total Focus on the Self?," in *The Lost Self: Pathologies of the Brain and Identity,* ed. Todd E. Feinberg and Julian Paul Keenan (Oxford: Oxford University Press, 2005).

134 **These new and revised theories:** This isn't an exhaustive discussion of the cognitive theories of autism. For another particularly interesting theory, see Laurent Mottron et al., "Enhanced Perceptual Functioning in Autism: An Update, and Eight Principles of Autistic Perception," *Journal of Autism and Developmental Disorders* 36, no. 1 (2006): 27–43.

Chapter 8: Another Path to Prodigy

135 **Though it's highly heritable:** See, for example, Sven Sandin et al., "The Familial Risks of Autism," *Journal of the American Medical Association* 311, no. 17 (2014): 1770–77.

136 **Similarly, some studies:** For a discussion of environmental factors potentially linked to autism, see Rodney R. Dietert, Janice M. Dietert, and Jamie C. Dewitt, "Environmental Risk Factors for Autism," *Emerging Health Threats Journal* 4 (2011).

136 **Some scientists have proposed:** Pauline Chaste and Marion Leboyer, "Autism Risk Factors: Genes, Environment, and Gene-Environment Interactions," *Dialogues in Clinical Neuroscience* 14, no. 3 (2012): 281–92. For a specific example of a potential gene-environment interaction, see Heather E. Volk et al., "Interaction of the MET Receptor Tyrosine Kinase Gene and Air Pollution Exposure in Autism Spectrum Disorder," *Epidemiology* 25, no. 1 (2014): 44–47.

137 **Josh Tiessen was born in Russia:** The events in this chapter described by Julie Tiessen come from telephone interviews conducted on Oct. 14 and 15, 2014, and June 10, 2015 (with occasional input from the rest of the family); and e-mail. In addition, Julie provided photographs and a written summary of Josh's development. The events in this chapter described by Valerie Jones come from a telephone interview conducted on Jan. 17, 2015. Josh's story was also drawn from his Web site; his article, "The Gift of Art Can Make the World a Better Place," *International Student Leaders: Students Globally, Learning to Lead,* ed. Ken Swan (Doctrina Education, Issue 2, 2014); and various news reports, including Alyice Edrich, "An Interview with Teen Emerging Artist, Josh Tiessen," EmptyEasel.com, Sept. 19, 2012; Conrad Collaco, "Stoney Creek Teen's Painting Hangs in the National Gallery," CBC News, June 1, 2012; Barrie Doyle, "Teen Artist Gets Advice from Canadian Painter," *Christian Week,* Sept. 2010; "Young Artist's Works Inspired by His Travels Around the World," *Burlington Post,* April 13, 2007; Leonard Turnevicius, "Sit Back and Watch the Music Come to

Life," *Hamilton Spectator,* July 4, 2013; Sarah Murrell, "An Art Prodigy," *Voice of Pelham,* May 1, 2013; Tyler Fyfe, "Painting Change—Joshua Tiessen," FreshPrintMagazine.com, March 3, 2014; and "Stoney Creek Studio Stroll Returns," *Stoney Creek News,* May 30, 2013.

140 **Art Gallery of Burlington:** At the time, it was called the Burlington Art Centre.

141 **Josh got his big break:** Robert Bateman to Josh Tiessen, Feb. 1, 2010, e-mail, provided by Julie Tiessen.

141 **Doug had been diagnosed:** For more information about the Tiessens' medical situation, see Jeff Mahoney, "Stoney Creek Family All Diagnosed with Lyme Disease," *Hamilton Spectator,* Aug. 29, 2014.

143 **Zac's childhood looked:** The details of Zac's development come from Julie Tiessen and Valerie Jones, as well as photographs, a written summary of Zac's development, and medical reports provided by Julie. Zac's story was also drawn from his Web site and various news reports, including Cory Ruf, "Identified as Arts Prodigies, Ontario Brothers Now Contributing to Science," CBC News, March 8, 2013; Laura Lennie, "Musical Blessings," *Stoney Creek News,* Dec. 11, 2013; and Joanna Frketich, "Child Prodigies to Take Part in Innovation Conference," *Hamilton Spectator,* July 4, 2013.

150 **It wasn't until Joanne:** For more on how the Tiessens and Joanne connected Zac's concussion with his music abilities, see Ruf, "Identified as Art Prodigies."

152 **"Please fix that":** David Henry Feldman, "Child Prodigies: Ancient Tradition, Recent Research" (presentation, New York University Langone Medical Center, Child Study Center, New York, April 5, 2012), slides provided by Feldman. Additional information on Feldman's NYU talk and grant application come from a telephone interview with him conducted on June 26, 2014; and e-mail.

152 **These individuals are known:** Treffert, "Savant Registry." In an April 2, 2015, telephone interview, Treffert said he thought that the 10 percent figure might actually be conservative.

152 **He envisioned pi:** For Jason's take on his story, see Jason Padgett and Maureen Seaberg, *Struck by Genius* (Boston: Houghton Mifflin Harcourt, 2014).

152 **Alonzo Clemons fell:** William E. Schmidt, "Gifted Retardates: The Search for Clues to Mysterious Talent," *New York Times,* July 12, 1983; Treffert, *Islands of Genius.* To see Alonzo sculpting, see *Ingenious Minds,* "Sculpting Prodigy," Science Channel; and *Beautiful Minds.*

153 **"as his personality crying":** Rob Lammie, "The Amazing Stories of 6 Sudden Savants," *Mental Floss,* June 29, 2010. For more on Tommy McHugh, see "Ex–Street Fighter, 60, Turned into a Fanatical Artist by a Brain

Haemorrhage That Physically Altered His Mind," *Daily Mail,* March 15, 2010; Helen Thomson, "Mindscapes: Stroke Turned Ex-Con into Rhyming Painter," *New Scientist,* May 10, 2013; and "Creative Side Unlocked by Stroke," BBC News, June 21, 2004.

153 **he also displayed impressive:** T. L. Brink, "Idiot Savant with Unusual Mechanical Ability: An Organic Explanation," *American Journal of Psychiatry* 137, no. 2 (1980): 250–51.

154 **Others had already observed:** Rimland, "Inside the Mind of the Autistic Savant."

154 **A psychologist examining:** Brink, "Idiot Savant with Unusual Mechanical Ability," 251.

154 **It was a line:** Bruce L. Miller et al., "Enhanced Artistic Creativity with Temporal Lobe Degeneration," *Lancet* 348, no. 904 (1996): 1744–45. For later, related works, see Bruce L. Miller et al., "Emergence of Artistic Talent in Frontotemporal Dementia," *Neurology* 51, no. 4 (1998): 978–82; and Bruce L. Miller et al., "Functional Correlates of Musical and Visual Ability in Frontotemporal Dementia," *British Journal of Psychiatry* 176, no. 5 (2000): 458–63.

154 **One such individual:** For more on this man, see Sheri Fink, "The Search for the Origins of Humankind's Creativity," *Oregonian,* July 1, 1998; and Graham Phillips, "The Scourge of Genius," *Sunday Telegraph,* Feb. 15, 1998.

154 **Miller and his colleagues eventually:** Miller et al., "Functional Correlates of Musical and Visual Ability in Frontotemporal Dementia."

154 **"that somehow a disease":** Rob Stein, "Patients' New Gift Paints Clearer Image of Disease," *Washington Post,* Oct. 26, 1998.

154 **Miller and his colleagues discovered:** Miller et al., "Functional Correlates of Musical and Visual Ability in Frontotemporal Dementia."

154 **Researchers put forward a flurry:** For an overview of these theories, see Darold A. Treffert and Daniel D. Christensen, "Inside the Mind of a Savant," *Scientific American,* Dec. 2005.

155 **Allan W. Snyder:** Information in this chapter on the experiments in which Allan Snyder and his colleagues attempted to induce skills comes from a telephone interview with Allan Snyder conducted on June 29, 2015; and academic articles, including Allan W. Snyder et al., "Savant-Like Skills Exposed in Normal People by Suppressing the Left Fronto-temporal Lobe," *Journal of Integrative Neuroscience* 2, no. 2 (2003): 149–58; Jason Gallate et al., "Reducing False Memories by Magnetic Pulse Stimulation," *Neuroscience Letters* 449 (2009): 151–54; Allan Snyder, "Savant-Like Numerosity Skills Revealed in Normal People by Magnetic Pulses," *Perception* 35 (2006): 837–45; Allan Snyder, "Explaining and Inducing Savant Skills: Privileged Access to Lower Level, Less-Processed Information," *Philosoph-*

ical Transactions of the Royal Society B 364 (2009): 1399–405; and Richard P. Chi and Allan W. Snyder, "Brain Stimulation Enables the Solution of an Inherently Difficult Problem," *Neuroscience Letters* 515, no. 2 (2012): 121–24.

155 **the classic example:** Oliver Sacks, *The Man Who Mistook His Wife for a Hat* (New York: Summit Books, 1985).

156 **Or maybe that inner savant:** For discussion of this point, see Robyn L. Young, Michael C. Ridding, and Tracy L. Morrell, "Switching Skills On by Turning Off Part of the Brain," *Neurocase* 10, no. 3 (2004): 215–22.

156 **But a 2004 study:** Ibid.

156 **in one other small study:** Chi and Snyder, "Brain Stimulation Enables the Solution of an Inherently Difficult Problem."

157 **After all, the ratio:** Treffert, *Islands of Genius*. A preliminary report generated using Treffert's savant registry found that the gender breakdown among acquired savants is also lopsided. See Treffert, "Savant Registry."

157 **There's a similar gender breakdown:** Among Bruce Miller's frontotemporal patients who gained skills following the onset of the disease, five of the seven were men (the split was closer to even among patients who maintained visual or musical abilities they had before the onset of dementia). See Miller et al., "Functional Correlates of Musical and Visual Ability in Frontotemporal Dementia."

157 **Studies have found:** See, for example, S. Baron-Cohen et al., "Elevated Fetal Steroidogenic Activity in Autism," *Molecular Psychiatry* 20, no. 3 (2015): 369–76.

157 **"perhaps surprising":** Berit Brogaard, Simo Vanni, and Juha Silvanto, "Seeing Mathematics: Perceptual Experience and Brain Activity in Acquired Synesthesia," *Neurocase* 19, no. 6 (2013): 566–75.

157 **There have been similarly puzzling:** Nathalie Boddaert et al., "Autism: Functional Brain Mapping of Exceptional Calendar Capacity," *British Journal of Psychiatry* 187, no. 1 (2005): 83–86.

157 **An fMRI of George Widener:** This fMRI was conducted as part of a PBS program, not an academic study. See "Mystery of the Savant Brain," *Nova,* transcript of episode that aired Oct. 24, 2012, quoting Joy Hirsch.

157 **The savants who demonstrated:** For a discussion of why this pattern might arise in math savants, see Brogaard, Vanni, and Silvanto, "Seeing Mathematics."

Chapter 9: Lightning in a Bottle

159 **As a toddler, Autumn:** The events in this chapter described by Autumn de Forest come from a telephone interview conducted on Oct. 30, 2014. The events in this chapter described by Doug de Forest come from telephone

interviews conducted on Nov. 19, 2014, and April 7 and Sept. 7, 2015; and e-mail. Autumn's story was also drawn from her Web site, marketing materials, and news reports, including "Could Most Modern Art Be Done by an 8-Year-Old? This Child Prodigy Proves That It Can!," *Daily Mail,* Oct. 15, 2010; "Pint-Size Picasso," *Time for Kids,* Aug. 30, 2013; "Autumn de Forest Interview: Young Prodigy Artist Inspires and Gives Back," TeensWan naKnow.com, June 7, 2014; Veronica, "Autumn de Forest on Artistic Inspiration!," SweetyHigh.com, April 23, 2014; Hugo Kugiya, "She's Just 8, yet She's Painted Art Worth $250,000," *Today,* Oct. 13, 2010; Bailey Powell, "An Interview with Child Prodigy Autumn de Forest," *Fort Worth Key Magazine,* Sept. 29, 2012; Terri Bryce Reeves, "Prodigy Autumn de Forest Is Latest Artist in de Forest Family," *Tampa Bay Times,* Nov. 8, 2012; Camille Moore and Brittany Taylor, "11-Year-Old Painter Autumn de Forest Spills What It's Like to Be an Art Prodigy," *Girls' Life,* July 4, 2013; Stephanie Anderson Witmer, "Ordinary Children, Extraordinary Talents," *USA Today Back to School,* Fall 2014; Erika Pope, "Empowering Autumn," *Vegas Seven,* Nov. 11, 2010; "Seven-Year-Old Artist Expected to Draw Attention at This Year's Malibu Chamber Arts Festival," *Malibu Surfside News,* July 16, 2009; Ben Marcus, "The Malibu Arts Festival Sets Up Camp at the Civic Center This Weekend," *Malibu Times,* July 22, 2009; Erica Tempesta, "Child Prodigy Autumn de Forest on Painting and Being One of Aéropostale's Epic Kids," *Styleite,* Feb. 17, 2014; "10 Art Prodigies You Should Know," *Huffington Post,* July 27, 2012; Julia Halperin, "From the Palettes of Babes: Four Prodigious Child Artists to Watch," *Huffington Post,* Jan. 31, 2011; and Autumn's television appearances, including on *Home & Family, One on One with Steve Adubato,* Discovery Health, *Studio 10, Inside Edition, The Wendy Williams Show,* and *Daytime.*

159 **"little kid drawings"**: Autumn de Forest, telephone interview, Oct. 30, 2014.

160 **"Like a Rothko"**: Reeves, "Prodigy Autumn de Forest Is Latest Artist in de Forest Family."

161 **"You see a spark"**: Witmer, "Ordinary Children, Extraordinary Talents."

161 **"Sometimes I wish"**: Autumn de Forest, Autumn de Forest Art Fans, Facebook, June 11, 2012.

163 **"fancy deer"**: Autumn's words at the time as recalled by Doug de Forest.

164 **"My spirit was on fire"**: Moore and Taylor, "11-Year-Old Painter Autumn de Forest Spills What It's Like to Be an Art Prodigy."

164 **She sold *Paradise*:** Information about Autumn's art sales provided by Doug de Forest.

165 **"an old lady in a young"**: *The Wendy Williams Show,* backstage conversation, clip available on Autumn's Web site.

165 **"a pistol"**: "Eight-Year-Old Girl Dazzles Art World," *Today,* Oct. 14, 2010.

166 **In one interview:** Tempesta, "Child Prodigy Autumn de Forest on Painting and Being One of Aéropostale's Epic Kids."

167 **The prodigies' average overall IQ score:** For more information on the
prodigies' cognitive profiles, see Joanne Ruthsatz, Kimberly Ruthsatz-
Stephens, and Kyle Ruthsatz, "The Cognitive Bases of Exceptional Abilities in Child Prodigies by Domain: Similarities and Differences,"
Intelligence 44 (2014): 11–14.

167 **There was only one real:** If you drop that child's score, the range for
working memory was 132–158.

168 **fluid reasoning:** "Frequently Asked Questions About the Stanford-Binet
Intelligence Scales, Fifth Edition," Thomson Nelson.

168 **Same story with:** For more information on what each of these subtests is
intended to measure, see Joel W. Schneider and Kevin S. McGrew, "The
Cattell-Horn-Carroll Model of Intelligence," in *Contemporary Intellectual
Assessment,* 3rd ed., ed. Dawn P. Flanagan and Patti L. Harrison (New
York: Guilford Press, 2012); and Dawn P. Flanagan and Shauna G. Dixon,
"The Cattell-Horn-Carroll Theory of Cognitive Abilities," in *Encyclopedia
of Special Education,* ed. Cecil R. Reynolds, Kimberly J. Vannest, and
Elaine Fletcher-Janzen (Hoboken, N.J.: John Wiley & Sons, 2013). For
information as to which subtests are rooted in which Cattell-Horn-Carroll
cognitive abilities, see Henry L. Janzen, John E. Obrzut, and Christopher
W. Marusiak, "Test Review: Roid, G. H. (2003). Stanford-Binet Intelligence
Scales, Fifth Edition (SB:V). Itasca, Ill.: Riverside Publishing," *Canadian
Journal of School Psychology* 19, nos. 1 and 2 (Dec. 2004): 235–44.

168 **prodigious skill in science:** Feldman and Morelock, "Prodigies and
Savants."

168 **They had an average score:** Some researchers have suggested that musical
practice may improve working memory. See, for example, Sissela Berg-
man Nutley, Fahimeh Darki, and Torkel Klingberg, "Music Practice Is
Associated with Development of Working Memory During Childhood
and Adolescence," *Frontiers in Human Neuroscience* 7 (2014).

168 **"seeing with the mind's eye":** For more on mental imagery, see Stephen
Michael Kosslyn, *Image and Brain: The Resolution of the Imagery
Debate* (Cambridge, Mass.: MIT Press, 1996).

168 **A *spatial visualizer's* mental imagery:** Some argue that spatial visualization
should be further broken down into spatial location and mental transformation. See William L. Thompson et al., "Two Forms of Spatial Imagery: Neu-
roimaging Evidence," *Psychological Science* 20, no. 10 (2009): 1245–53.

168 **A 1985 study:** David N. Levine, Joshua Warach, and Martha Farah, "Two
Visual Systems in Mental Imagery: Dissociation of 'What' and 'Where' in
Imagery Disorders Due to Bilateral Posterior Cerebral Lesions," *Neurol-
ogy* 35, no. 7 (1985): 1010–18.

169 **The Stanford-Binet:** Janzen, Obrzut, and Marusiak, "Test Review: Roid, G. H. (2003)."

169 **"ability to perceive complex":** Schneider and McGrew, "Cattell-Horn-Carroll Model of Intelligence," 129. See also Flanagan and Dixon, "The Cattell-Horn-Carroll Theory of Cognitive Abilities."

169 **This type of visualization:** See, for example, Maria Kozhevnikov et al., "Creativity, Visualization Abilities, and Visual Cognitive Style," *British Journal of Educational Psychology* 83, no. 2 (2013): 196–209.

169 **Consider, for example:** "Groupon, Qatar, Jake Barnett."

169 **Galileo similarly visualized:** Arthur I. Miller, *Insights of Genius: Imagery and Creativity in Science and Art* (Cambridge, Mass.: MIT Press, 2000).

169 **In theory, this left object:** Olesya Blazhenkova and Maria Kozhevnikov have advocated for recognizing object visualization as a component of intelligence. See "Visual-Object Ability: A New Dimension of Non-Verbal Intelligence," *Cognition* 117, no. 3 (2010): 276–301.

169 **a skill tied to artistic ability:** See, for example, Kozhevnikov et al., "Creativity, Visualization Abilities, and Visual Cognitive Style."

170 **the price of excelling at one:** Maria Kozhevnikov, Olesya Blazhenkova, and Michael Becker, "Trade-Off in Object Versus Spatial Visualization Abilities: Restriction in the Development of Visual-Processing Resources," *Psychonomic Bulletin and Review* 17, no. 1 (2010): 29–35.

171 **"I just hear it":** "Musical Prodigy, Bluejay," *60 Minutes.* For more information on Jay, see Matthew Gurewitsch, "Early Works of a New Composer (Very Early, in Fact)," *New York Times*, Aug. 13, 2006.

171 **Jacob Barnett, the science prodigy:** Kristine Barnett, *Spark.*

171 **The music prodigy Jonathan:** Eve Weiss, telephone interview, Jan. 20, 2014.

171 **The object and spatial visualization abilities:** For a brief overview, see Kozhevnikov, Blazhenkova, and Becker, "Trade-Off in Object Versus Spatial Visualization Abilities." See also Levine, Warach, and Farah, "Two Visual Systems in Mental Imagery."

172 **Over several decades, researchers:** Scientists have sought out abnormalities in brain structure and functioning as well. We are focusing here on those factors that have also been investigated in prodigies—behaviors, cognitive tendencies, and genetics. But for a paper arguing that attempts to find a brain abnormality shared by all autists have failed, see Lynn Waterhouse and Christopher Gillberg, "Why Autism Must Be Taken Apart," *Journal of Autism and Developmental Disorders* 44, no. 7 (2014): 1788–92.

172 **"the children's *inability to relate*":** Kanner, "Autistic Disturbances of Affective Contact," 242.

172   **But in the early years:** For a review of the varying autism criteria used by early investigators, see Rutter, "Diagnosis and Definition of Childhood Autism."

172   **By the early 1970s:** Michael Rutter, "Autistic Children: Infancy to Adulthood," *Seminars in Psychiatry* 2, no. 4 (1970): 435–50.

173   **"into question the usefulness":** Wing and Gould, "Severe Impairments of Social Interaction and Associated Abnormalities in Children," 27.

173   **Recently, a team:** Francesca Happé and Angelica Ronald, "The 'Fractionable Autism Triad': A Review of Evidence from Behavioural, Genetic, Cognitive, and Neural Research," *Neuropsychology Review* 18, no. 4 (2008): 287–304.

173   **Even autistic *siblings*:** Ryan K. C. Yuen et al., "Whole-Genome Sequencing of Quartet Families with Autism Spectrum Disorder," *Nature Medicine* 21, no. 2 (2015): 185–91.

173   **"if you've met one person with autism":** This quotation is often attributed to Stephen Shore, a clinical assistant professor at Adelphi University who frequently speaks and writes about autism.

173   **Researchers struggled mightily:** For a brief review, see Francesca Happé, Angelica Ronald, and Robert Plomin, "Time to Give Up on a Single Explanation for Autism," *Nature Neuroscience* 9, no. 10 (2006): 1218–20.

173   **Initial optimism:** For a discussion of the historical perspective, see Judith H. Miles, "Autism Spectrum Disorders—a Genetics Review," *Genetics in Medicine* 13, no. 4 (2011): 278–94.

173   **Researchers found not one:** For a review, see Jamee M. Berg and Daniel H. Geschwind, "Autism Genetics: Searching for Specificity and Convergence," *Genome Biology* 13, no. 7 (2012): 247.

173   **Even the most prevalent:** Shafali S. Jeste and Daniel H. Geschwind, "Disentangling the Heterogeneity of Autism Spectrum Disorder Through Genetic Findings," *Nature* 10, no. 2 (2014): 74–81.

173   **"the genetic architecture":** D. Q. Ma et al., "A Genome-Wide Association Study of Autism Reveals a Common Novel Risk Locus at 5p14.1," *Annals of Human Genetics* 73, no. 3 (2009): 268–73, 270.

173   **It turns out that even:** Yuen et al., "Whole-Genome Sequencing of Quartet Families with Autism Spectrum Disorder."

173   **This heterogeneity of behaviors:** For two examples, see Eric B. London, "Categorical Diagnosis: A Fatal Flaw for Autism Research?," *Trends in Neurosciences* 37, no. 12 (2014): 683–86; and Happé, Ronald, and Plomin, "Time to Give Up on a Single Explanation for Autism."

174   **He and his team have tied:** See, for example, Philip Awadalla et al., "Direct Measure of the De Novo Mutation Rate in Autism and Schizophrenia Cohorts," *American Journal of Human Genetics* 87, no. 3 (2010): 316–24.

174   **It was this idea:** Guy Rouleau, telephone interview, Dec. 18, 2014.

174 **One group of researchers:** Nurmi et al., "Exploratory Subsetting of Autism Families Based on Savant Skills Improves Evidence of Genetic Linkage to 15q11-q13."

174 **But when another team:** Ma et al., "Ordered-Subset Analysis of Savant Skills in Autism for 15q11-q13."

Chapter 10: The Recovery Enigma

176 **Her older son, Alex:** Lucie, telephone interview, Sept. 4, 2014; Alex, Grade 5 Report Card, June 19, 2015.

176 **Her second son, William:** Lucie, telephone interview, Sept. 12, 2014; and e-mail; Josh (William's math teacher), telephone interview, Oct. 14, 2014; William, Grade 4 Report Card, June 19, 2015.

177 **patient who no longer seemed autistic:** See, for example, Kanner and Eisenberg, "Notes on the Follow-Up Studies of Autistic Children." The authors include a description of Robert F., an individual characterized as having reached a "higher pinnacle" than the rest. He had served in the navy, married, and was studying musical composition. Still, the authors cautioned that in most cases "emergence" from autism was only "partial."

177 **"somewhat odd":** Eisenberg, "Autistic Child in Adolescence," 608.

177 **One study described:** Rutter, Greenfeld, and Lockyer, "A Five to Fifteen Year Follow-Up Study of Infantile Psychosis."

177 **there was a report:** Rutter, "Autistic Children." See also Marian K. DeMyer et al., "Prognosis in Autism: A Follow-Up Study," *Journal of Autism and Childhood Schizophrenia* 3, no. 3 (1973): 233 (describing "two autistic children who were 'normal' in all respects at follow-up"); Janet L. Brown, "Adolescent Development of Children with Infantile Psychosis," *Seminars in Psychiatry* 1 (1969): 79–89 (identifying some autistic children eventually deemed "normal," though noting that "the social development of even the best-functioning of these children will be markedly retarded and that each step is accomplished with painful difficulty").

177 **Many thought autism:** For an overview of recovery-related research, see Molly Helt et al., "Can Children with Autism Recover? If So, How?," *Neuropsychology Review* 18, no. 4 (2008): 339–66.

177 **She and her team:** Deborah Fein et al., "Optimal Outcome in Individuals with a History of Autism," *Journal of Child Psychology and Psychiatry* 54, no. 2 (2013): 195–205. This study was a continuation of work that Fein and her team had started earlier. See Elizabeth Kelley et al., "Residual Language Deficits in Optimal Outcome Children with a History of Autism," *Journal of Autism and Developmental Disorders* 36, no. 6 (2006): 807–28; Elizabeth Kelley, Letitia Naigles, and Deborah Fein, "An In-Depth Examination of Optimal Outcome Children with a History of

Autism Spectrum Disorders," *Research in Autism Spectrum Disorders* 4, no. 3 (2010): 526–38. Additional information on Fein's work comes from a telephone interview with Deborah Fein conducted on March 2, 2015.

177 **They didn't have any residual:** Eva Troyb et al., "Academic Abilities in Children and Adolescents with a History of Autism Spectrum Disorders Who Have Achieved Optimal Outcomes," *Autism* 18, no. 3 (2014): 233–43.

178 **They cautioned that an optimal:** Fein et al., "Optimal Outcome in Individuals with a History of Autism."

178 **"the 'r' word":** Sally Ozonoff, "Editorial: Recovery from Autism Spectrum Disorder (ASD) and the Science of Hope," *Journal of Child Psychology and Psychiatry* 54, no. 2 (2013): 113–14, 114.

178 **"Symptoms alone rarely indicate":** Thomas Insel, "Director's Blog: Transforming Diagnosis," National Institute of Mental Health, April 29, 2013, http://www.nimh.nih.gov/.

179 **growing consensus:** See, for example, Rutter, "Autistic Children."

179 **In one of the earliest experiments:** C. B. Ferster and Marian K. DeMyer, "The Development of Performances in Autistic Children in an Automatically Controlled Environment," *Journal of Chronic Diseases* 13, no. 4 (1961): 312–45.

179 **Two years later, another team:** Montrose Wolf, Todd Risley, and Hayden Mees, "Application of Operant Conditioning Procedures to the Behaviour Problems of an Autistic Child," *Behaviour Research and Therapy* 1, nos. 2–4 (1963): 305–12.

179 **In contrast to the failure:** Laura Schreibman, "Intensive Behavioral/Psychoeducational Treatments for Autism: Research Needs and Future Directions," *Journal of Autism and Developmental Disorders* 30, no. 5 (2000): 373–78.

179 **In the early 1970s:** O. Ivar Lovaas et al., "Some Generalization and Follow-Up Measures on Autistic Children in Behavior Therapy," *Journal of Applied Behavior Analysis* 6, no. 1 (1973): 131–66.

179 **In a follow-up investigation:** O. Ivar Lovaas, "Behavioral Treatment and Normal Educational and Intellectual Functioning in Young Autistic Children," *Journal of Consulting and Clinical Psychology* 55, no. 1 (1987): 3–9.

179 **A few years later:** John J. McEachin, Tristram Smith, and O. Ivar Lovaas, "Long-Term Outcome for Children with Autism Who Received Early Intensive Behavioral Treatment," *American Journal on Mental Retardation* 97, no. 4 (1993): 359–72.

179 **Some scientists suggested:** Eric Schopler, Andrew Short, and Gary Mesibov, "Relation of Behavioral Treatment to 'Normal Functioning': Comment on Lovaas," *Journal of Consulting and Clinical Psychology* 57, no. 1 (1989):

162–64; Peter Mundy, "Normal Versus High-Functioning Status in Children with Autism," *American Journal on Mental Retardation* 97, no. 4 (1993): 381–84.

180 **They view efforts to eradicate:** See, for example, "Position Statements," Autistic Self Advocacy Network, http://autisticadvocacy.org/policy-advo cacy/position-statements.

180 **recognize the unique contributions:** Laurent Mottron, telephone interview, Aug. 16, 2015.

180 **In terms of effectiveness:** Amy S. Weitlauf et al. (the Vanderbilt Evidence-Based Practice Center), *Therapies for Children with Autism Spectrum Disorder: Behavioral Interventions Update,* Agency for Healthcare Research and Quality, *Comparative Effectiveness Review* 137 (2014): 80.

180 **a 2015 review:** Brian Reichow, "Overview of Meta-Analyses on Early Intensive Behavioral Intervention for Young Children with Autism Spectrum Disorders," *Journal of Autism and Developmental Disorders* 42, no. 4 (2012): 512–20, 518.

180 **the U.S. surgeon general:** U.S. Public Health Service, Office of the Surgeon General, *Mental Health: A Report of the Surgeon General* (National Institute of Mental Health, 1999).

180 **Autism Speaks:** "Treatments & Therapies," Autism Speaks, https://www .autismspeaks.org.

180 **Not every kid who receives:** Patricia Howlin, Iliana Magiati, and Tony Charman, "Systematic Review of Early Intensive Behavioral Interventions for Children with Autism," *American Journal on Intellectual and Developmental Disabilities* 114, no. 1 (2009): 23–41.

181 **But there were still optimal outcome kids:** Alyssa J. Orinstein et al., "Intervention for Optimal Outcome in Children and Adolescents with a History of Autism," *Journal of Developmental and Behavioral Pediatrics* 35, no. 4 (2014): 247–56.

181 **Almost from birth, Ping Lian Yeak:** The events in this chapter described by Sarah Lee come from e-mail. In addition, Sarah provided photographs, a transcription of Ping Lian's speech-language assessment dated Nov. 2, 1997, excerpts from her journal, and portions of her manuscript about raising Ping Lian. The events in this chapter described by Rosa C. Martinez come from a telephone interview conducted on March 5, 2015. The events in this chapter described by Laurence A. Becker come from a telephone interview conducted on Dec. 17, 2014. Ping Lian's story was also drawn from his Web site, marketing materials, and various news reports, including Mark White, "Island of Genius," *Sydney Morning Herald,* April 12, 2014; Tan Sher Lynn, "When Love, Hope & Faith Endure," *KL Lifestyle,* Jan. 2011; Shanti Ganesan, "Fate & Destiny," *Marie Claire Malaysia,* May 2008; Angus Fontaine, "Ping Lian Yeak," *Time Out Sydney,* Sept. 19, 2011; "The World Is His

Canvas," *Passions,* Sept. 2009; Koh Soo Ling, "The Road Less Travelled," *New Straits Times,* Jan. 16, 2005; Vivienne Pal, "Strokes of Genius from an 11-Year-Old Autistic Child," *Star,* Feb. 3, 2005; Ruth Wong, "Through the Eyes of Love," *Asia!,* May 17, 2009; "Brilliant Art of an Autistic Child," *New Straits Times,* June 21, 2005; Arni Shahida Razak, "At the Art of Autism," *New Straits Times,* Sept. 26, 2004; Andrew Priestley, "World's Eye on Autistic Artist," *North Shore Times,* Jan. 23, 2009; Barbara Foong, "Different Strokes by Special Needs Persons," *New Straits Times,* Dec. 10, 2003; Jenny Hatton Mahon, "Autism—The Art of Autism," Weekendnotes.com, July 9, 2014; Jessica Lim, "Capturing Genius on Film," *New Straits Times,* Sept. 11, 2005; and Ping Lian's television appearance on SBS News.

182 **"Ping Lian presents with moderate":** Speech-language assessment, Nov. 2, 1997, document transcribed and provided by Sarah Lee.

183 **"I tell myself":** Sarah Lee, journal entry, Feb. 2004, transcribed and provided by Sarah Lee.

184 **"totally focused and full of energy":** Sarah Lee, excerpt from draft manuscript of book on raising Ping Lian, *"I Want to Be an Artist": An Autistic Savant's Voice and a Mother's Dream Transformed onto Canvas.*

184 **"anywhere and everywhere":** Ibid.

184 **"He seemed almost obsessed":** Ibid.

185 **"deep in the eye":** Fontaine, "Ping Lian Yeak."

186 **One of his pieces:** Dollar amount converted from Malaysian ringgit.

187 **"imposing in their intricacy":** White, "Island of Genius."

187 **"vivid splashes of color":** Ibid.

187 **"bold strokes and cheerful colours":** Wong, "Through the Eyes of Love."

188 **"Great artist":** "SBS News—Savant Artist in Spotlight," YouTube video, 4:14, posted by "WorldNews Australia," Sept. 28, 2011.

188 **"become so meaningful":** Sarah Lee, e-mail.

188 **Training the talent:** There are several places focused on developing autistic strengths, such as the Tailor Institute in Missouri, Hidden Wings in California, and Strokes of Genius in New York.

189 **It's an intriguing approach:** A researcher in Australia, Trevor Clark, has been working on developing a train-the-talent curriculum; the results of his work are as yet unpublished. For a description, see Trevor Clark, "The Application of Savant and Splinter Skills in the Autistic Population Through an Educational Curriculum," Wisconsin Medical Society, https://www.wisconsinmedicalsociety.org/.

189 **newer models of behavioral therapy:** For an overview of more play-based, child-driven approaches, see Laura Schreibman et al., "Naturalistic Developmental Behavioral Interventions: Empirically Validated Treatment for Autism Spectrum Disorder," *Journal of Autism and Developmental Disorders* 45, no. 8 (2015): 2411–28.

189   **though a tendency toward obsession is a widely recognized:** See, for example, Klin et al., "Circumscribed Interests in Higher Functioning Individuals with Autism Spectrum Disorders."

189   **Kanner, for example, questioned:** Kanner, "Autistic Disturbances of Affective Contact," 243.

189   **More recently, scientists considering:** Richard C. Barnes and Stephen M. Earnshaw, "Problems with the Savant Syndrome: A Brief Case Study," *British Journal of Learning Disabilities* 23, no. 3 (1995): 124–26.

189   **There's evidence that, in contrast:** Klin et al., "Circumscribed Interests in Higher Functioning Individuals with Autism Spectrum Disorders"; Mercier, Mottron, and Belleville, "Psychosocial Study on Restricted Interests in High-Functioning Persons with Pervasive Developmental Disorders"; Turner-Brown et al., "Phenomenology and Measurement of Circumscribed Interests in Autism Spectrum Disorders."

190   **Several small studies:** Mary J. Baker, Robert L. Koegel, and Lynn Kern Koegel, "Increasing the Social Behavior of Young Children with Autism Using Their Obsessive Behaviors," *Research and Practice for Persons with Severe Disabilities* 23, no. 4 (1998): 300–308; Mary J. Baker, "Incorporating the Thematic Ritualistic Behaviors of Children with Autism into Games: Increasing Social Play Interactions with Siblings," *Journal of Positive Behavior Interventions* 2, no. 2 (2000): 66–84.

190   **others have found:** Marjorie H. Charlop-Christy and Linda K. Haymes, "Using Objects of Obsession as Token Reinforcers for Children with Autism," *Journal of Autism and Developmental Disorders* 28, no. 3 (1998): 189–98; Marjorie H. Charlop, Patricia F. Kurtz, and Fran Greenberg Casey, "Using Aberrant Behaviors as Reinforcers for Autistic Children," *Journal of Applied Behavior Analysis* 23, no. 2 (1990): 163–81; Marjorie H. Charlop-Christy and Linda K. Haymes, "Using Obsessions as Reinforcers With and Without Mild Reductive Procedures to Decrease Inappropriate Behaviors of Children With Autism," *Journal of Autism and Developmental Disorders* 26, no. 5 (1996): 527–46; Laurie A. Vismara and Gregory L. Lyons, "Using Perseverative Interests to Elicit Joint Attention Behaviors in Young Children With Autism," *Journal of Positive Behavior Interventions* 9, no. 4 (2007): 214–28; Brian A. Boyd et al., "Effects of Circumscribed Interests on the Social Behaviors of Children with Autism Spectrum Disorders," *Journal of Autism and Developmental Disorders* 37, no. 8 (2007): 1550–61.

190   **In 1944, in his first published paper on autism:** Asperger, "'Autistic Psychopathy' in Childhood."

190   **"the two real success stories":** Kanner, "Follow-Up Study of Eleven Autistic Children Originally Reported in 1943," 143.

190   **The savant expert:** Treffert, *Islands of Genius*.

190 **Temple Grandin:** Temple Grandin and Kate Duffy, *Developing Talents: Careers for Individuals with Asperger Syndrome and High-Functioning Autism* (Shawnee Mission, Kans.: Autism Asperger Publishing, 2008).

191 **Until scientists parse out:** Stephen Bent and Robert L. Hendren, "Improving the Prediction of Response to Therapy in Autism," *Neurotherapeutics* 7, no. 3 (2010): 232–40; Happé, Ronald, and Plomin, "Time to Give Up on a Single Explanation for Autism."

192 **"The best way to better services":** Thomas Insel, "Director's Blog: Autism Awareness: April 2014," National Institute of Mental Health, March 27, 2014, http://www.nimh.nih.gov/.

Chapter 11: The Next Quest

193 **"divine inspiration":** Feldman and Morelock, "Prodigies and Savants," 214.

195 **Take Erich Fuchs and Stephen Crohn:** For more information on Erich and Stephen, see Jesse Green, "The Man Who Was Immune to AIDS," *New York,* June 13, 2014; Gina Kolata, "Sharing of Profits Is Debated as the Value of Tissue Rises," *New York Times,* May 15, 2000; Elaine Woo, "Stephen Crohn Dies at 66; Immune to HIV, but Not Its Tragedy," *Los Angeles Times,* Sept. 21, 2013; John Schwartz, "Stephen Crohn, Who Furthered AIDS Study, Dies at 66," *New York Times,* Sept. 14, 2013; and "Surviving AIDS," *Nova,* transcript of episode that aired Feb. 2, 1999.

195 **"It was clear the minute":** Bill Paxton, telephone interview, April 23, 2015.

195 **Paxton and his colleagues:** Information regarding this study comes from William A. Paxton et al., "Relative Resistance to HIV-1 Infection of CD4 Lymphocytes from Persons Who Remain Uninfected Despite Multiple High-Risk Sexual Exposures," *Nature Medicine* 2, no. 4 (1996): 412–17; and a telephone interview with Paxton conducted on April 23, 2015.

195 **The scientists used ever-increasing doses:** In a later study, it was reported that it took roughly a thousand times more virus to infect Erich's and Stephen's cells than it did to infect the control cells; even with that amount of exposure, the virus took hold in only a small fraction of cells, and it failed to replicate. See Rong Liu et al., "Homozygous Defect in HIV-1 Coreceptor Accounts for Resistance of Some Multiply-Exposed Individuals to HIV-1 Infection," *Cell* 86, no. 3 (1996): 367–77.

195 **"We repeated and repeated":** Bill Paxton, telephone interview, April 23, 2015.

196 **Across three studies:** Yaoxing Huang et al., "The Role of a Mutant CCR5 Allele in HIV-1 Transmission and Disease Progression," *Nature Medicine* 2, no. 11 (1996): 1240–43; Michael Dean et al., "Genetic Restriction of HIV-1 Infection and Progression to AIDS by a Deletion Allele of the CKR5 Structural Gene," *Science* 273, no. 5283 (1996): 1856–61; Michael Samson et al.,

"Resistance to HIV-1 Infection in Caucasian Individuals Bearing Mutant Alleles of the CCR-5 Chemokine Receptor Gene," *Nature* 382, no. 6593 (1996): 722–25.

197 **Timothy was a twenty-nine-year-old:** The events in this chapter described by Timothy Ray Brown come from telephone interviews conducted on April 3 and 22, 2015 (with occasional input from Dave Purdy). Events in this chapter described by Dr. Gero Hütter come from a telephone interview conducted on April 28, 2015. Timothy's story was also drawn from two articles he wrote: "I Am the Berlin Patient: A Personal Reflection," *AIDS Research and Human Retroviruses* 31, no. 1 (2015): 2–3, and "Cure: The Beginning of the End of HIV and AIDS," in *How AIDS Ends: An Anthology from San Francisco AIDS Foundation,* ed. Reilly O'Neal (New York: Vintage Books, 2004), excerpt printed in *POZ,* Nov. 26, 2012; news reports, including Tina Rosenberg, "The Man Who Had HIV and Now Does Not," *New York,* May 29, 2011; Regan Hofmann, "Patient No More," *POZ,* June 2011; Apoorva Mandavilli, "The AIDS Cure," *Popular Science,* March 7, 2014; and Mark Schoofs, "A Doctor, a Mutation, and a Potential Cure for AIDS," *Wall Street Journal,* Nov. 7, 2008.

200 **"functionally cured":** Schoofs, "A Doctor, a Mutation, and a Potential Cure for AIDS."

200 **The *New England Journal of Medicine:*** Gero Hütter et al., "Long-Term Control of HIV by CCR5 Delta32/Delta32 Stem-Cell Transplantation," *New England Journal of Medicine* 360, no. 7 (2009): 692–98.

200 **In a 2011 paper:** Kristina Allers et al., "Evidence for the Cure of HIV Infection by CCR5Δ32/Δ32 Stem Cell Transplantation," *Blood* 117, no. 10 (2011): 2791–99.

200 **Two months later:** "Is 'Cure' Still a Four-Letter Word? Executive Summary," San Francisco AIDS Foundation, http://www.sfaf.org.

200 **At a conference in Spain:** Richard Knox, "Traces of Virus in Man Cured of HIV Trigger Scientific Debate," NPR, June 13, 2012; Jon Cohen, "Evidence That Man Cured of HIV Harbors Viral Remnants Triggers Confusion," *Science,* June 11, 2012.

200 **So far, no second cure:** Lambros Kordelas et al., "Shift of HIV Tropism in Stem-Cell Transplantation with CCR5 Delta32 Mutation," *New England Journal of Medicine* 371, no. 9 (2014): 880–82; Gero Hütter, "More on Shift of HIV Tropism in Stem-Cell Transplantation with CCR5 Delta32/Delta32 Mutation," *New England Journal of Medicine* 371, no. 25 (2014): 2437–38.

201 **In 2014, a group:** Pablo Tebas et al., "Gene Editing of CCR5 in Autologous CD4 T Cells of Persons Infected with HIV," *New England Journal of Medicine* 370, no. 10 (2014): 901–10. Additional information on this study comes from a telephone interview with Pablo Tebas conducted on March 6, 2015.

202 **In one recent study:** Jason Flannick et al., "Loss-of-Function Mutations in SLC30A8 Protect Against Type 2 Diabetes," *Nature Genetics* 46, no. 4 (2014): 357–63. Additional information on this study comes from a telephone interview with Jason Flannick conducted on March 11, 2015.

202 **Scientists studying heart disease:** Jonathan Cohen et al., "Low LDL Cholesterol in Individuals of African Descent Resulting from Frequent Nonsense Mutations in PCSK9," *Nature Genetics* 37, no. 2 (2005): 161–65; Jonathan C. Cohen et al., "Sequence Variations in PCSK9, Low LDL, and Protection Against Coronary Heart Disease," *New England Journal of Medicine* 354, no. 12 (2006): 1264–72.

202 **The identification of beneficial:** Gina Kolata, "Rare Mutation Ignites Race for Cholesterol Drug," *New York Times,* July 9, 2013.

203 **Some advocates argue:** For an excellent history of this idea, which is often referred to as neurodiversity, see Silberman, *Neurotribes.* See also "Position Statements," Autistic Self Advocacy Network; Amy Harmon, "How About Not 'Curing' Us, Some Autistics Are Pleading," *New York Times,* Dec. 20, 2004; Amy Harmon, "Nominee to Disability Council Is Lightning Rod for Dispute on Views of Autism," *New York Times,* March 27, 2010; and Michelle Dawson, "The Misbehaviour of Behaviourists: Ethical Challenges to the Autism-ABA Industry," Jan. 18, 2004.

203 **"Autism isn't something a person *has*":** Jim Sinclair, "Don't Mourn for Us," *Our Voice* 1, no. 3 (1993).

203 **"The biggest barrier":** Julia Bascom, e-mail.

204 **It's a disorder:** Kit Weintraub, "A Mother's Perspective," Association for Science in Autism Treatment; Harmon, "How About Not 'Curing' Us."

204 **certain SLC30A8 mutations:** Flannick et al., "Loss-of-Function Mutations in SLC30A8 Protect Against Type 2 Diabetes." See also Gina Kolata, "Rare Mutation Kills Off Gene Responsible for Diabetes," *New York Times,* March 2, 2014.

204 **"It was so at odds":** Jason Flannick, telephone interview, March 11, 2015.

205 **"We're not really trying to cure":** Geraldine Dawson, telephone interview, Oct. 14, 2015.

206 **It described groups of symptoms:** Insel, "Director's Blog: Transforming Diagnosis."

206 **"As long as the research community":** Pam Belluck and Benedict Carey, "Psychiatry's Guide Is out of Touch with Science, Experts Say," *New York Times,* May 6, 2013.

206 **"It was all like a magical mystery tour":** Bruce Cuthbert, interview, March 9, 2015.

207 **It was even more problematic:** Linda S. Brady and Thomas R. Insel, "Translating Discoveries into Medicine: Psychiatric Drug Development in 2011," *Neuropsychopharmacology* 37, no. 1 (2012): 281–83; Bruce N. Cuthbert

and Thomas R. Insel, "Toward the Future of Psychiatric Diagnosis: The Seven Pillars of RDoC," *BMC Medicine* 11 (2013); Sten Stovall, "R&D Cuts Curb Brain-Drug Pipeline," *Wall Street Journal,* March 27, 2011.

207 **NIMH scientists eventually set out:** Cuthbert and Insel, "Toward the Future of Psychiatric Diagnosis."

207 **RDoC casts aside:** "Development and Definitions of the RDoC Domains and Constructs," National Institute of Mental Health, http://www.nimh .nih.gov/.

207 **The idea is that by untangling:** Insel, "Director's Blog: Transforming Diagnosis."

208 **But they are attempting:** Cuthbert and Insel, "Toward the Future of Psychiatric Diagnosis."

208 **After all, scientists have already:** Iva Dincheva et al., "FAAH Genetic Variation Enhances Fronto-amygdala Function in Mouse and Human," *Nature Communications* 6, no. 6395 (2015). See also Richard A. Friedman, "The Feel-Good Gene," *New York Times,* March 6, 2015.

Epilogue: A Wide-Open Future

209 **There are plenty of popular reports:** See, for example, Ellen Winner, "Often, Child Prodigies Do Not Grow into Adult Geniuses," *New York Times,* May 20, 2015; "Harvard's Quartet of Mental Prodigies," *New York Times,* Jan. 16, 1910; Kathleen Montour, "William James Sidis, the Broken Twig," *American Psychologist* 32, no. 4 (1977): 265–79; "Illustrating a System of Education," *New York Times,* Jan. 7, 1910.

209 **Just as autism in adulthood:** Siri Carpenter, "Adults with Autism Are Left to Navigate a Jarring World," *ScienceNews,* Feb. 10, 2015.

209 **wide range of outcomes:** For a thoughtful discussion on the transition from child prodigy to adult creator, see Ellen Winner, *Gifted Children: Myths and Realities* (New York: Basic Books, 1996). See also Michael J. A. Howe, *The Psychology of High Abilities* (New York: New York University Press, 1999).

209 **A small study of eight:** Robert W. Howard, "Linking Extreme Precocity and Adult Eminence: A Study of Eight Prodigies at International Chess," *High Ability Studies* 19, no. 2 (2008): 117–30.

209 **David Feldman has lost touch:** David Feldman, telephone interview, July 6, 2015.

210 **Pursuits that once came naturally:** Jeanne Bamberger, "Growing Up Prodigies: The Midlife Crisis," *New Directions for Child and Adolescent Development* 17 (1982): 61–77.

210 **Critics can seem almost gleeful:** John Radford, "Prodigies in the Press," *High Ability Studies* 9, no. 2 (1998): 153–64; Goode, "Uneasy Fit of the Precocious and the Average."

211 **Greg Grossman:** Greg Grossman, telephone interview, July 12, 2015.

211 **Jonathan Russell:** Eve Weiss, e-mail, July 3, 2015; Jonathan Russell, telephone interview, Jan. 20, 2014.

212 **Lauren Voiers:** Lauren Voiers, telephone interview, Sept. 8, 2015, and e-mail; Soeder, "Cleveland Artist Lauren Voiers Sculpts John Lennon Tribute for Liverpool Park."

212 **Richard Wawro:** Mike Wawro, telephone interview, Dec. 16, 2014.

212 **Jacob Barnett:** Kristine Barnett, telephone interview, Dec. 22, 2014, and *Spark;* Wells, "Jacob Barnett, Boy Genius."

213 **Afterward, he and his mentor:** Yogesh N. Joglekar and Jacob L. Barnett, "Origin of Maximal Symmetry Breaking in Even PT-Symmetric Lattices," *Physical Review A* 84 (2011).

213 **Jourdan Urbach:** Jourdan Urbach, interview, July 2, 2014; Todd Spangler, "Ocho, 8-Second Social-Video Startup, Raises $1.65 Million from Mark Cuban and Others," *Variety*, Nov. 11, 2014.

213 **the paper he had coauthored with Joanne:** Ruthsatz and Urbach, "Child Prodigy."

214 **Josh and Zac Tiessen:** Julie Tiessen, telephone interviews, June 10 and July 24, 2015, and e-mail; Cory Ruf, "Stoney Creek Painter Josh Tiessen One of Canada's 'Top 20 Under 20,'" CBC Hamilton, June 6, 2013; "Chronic Lyme Disease: Tragedy for Family of Four," YouCaring, postings dated Aug. 2014 to Sept. 2015.

214 **By the time he was nineteen:** Sales information provided by Julie Tiessen.

215 **"a blistering three song set":** Hunter Foley, "Review+Photos: Animals as Leaders w / After the Burial + Guests," *Heavy Press*, March 7, 2014.

215 **At eighteen:** Damian Fanelli, "Double-Hand Thumb Tapping with Eight-String Guitarist Zac Tiessen," *Guitar World*, Feb. 13, 2015.

215 **the magazine also recently spotlighted:** Jackson Maxwell, "Eight-String Guitarist Zac Tiessen Premieres 'Infinity' Playthrough Video," *Guitar World*, July 23, 2015.

216 **Autumn de Forest:** Doug de Forest, telephone interview, Sept. 7, 2015.

216 **"I've never taken lessons":** "Autumn, a Young Artist, Prodigy, and Philanthropist, Citizen Kid by Disney," YouTube video, 2:51, posted by "Babble," July 16, 2014.

216 **Alex and William:** Lucie, e-mail; Josh (William's math teacher), telephone interview, Oct. 14, 2014; William, report card, June 19, 2015.

Index

Note: Some names have been changed to protect the privacy of the individuals described.